Integrated Circuit Design

Alan F. Murray

and

H. Martin Reekie

Integrated Circuit Design

Springer Science+Business Media, LLC

© Alan F. Murray and H. Martin Reekie 1987
Originally published by Springer-Verlag New York Inc. in 1987
Softcover reprint of the hardcover 1st edition 1987

All rights reserved. No part of this
publication may be reproduced or transmitted,
in any form or by any means, without permission.

First published 1987 by
MACMILLAN EDUCATION LTD
London and Basingstoke

Library of Congress Cataloging-in-Publication Data
Murray, Alan F.
    Integrated circuit design.
    Bibliography: p.
    Includes index.
    1. Integrated circuits. 2. Electronic circuit
design. I. Reekie, H. Martin. II. Title
TK7874.M88   1987   621.381'73   87-12935

ISBN 978-1-4899-6677-3   ISBN 978-1-4899-6675-9 (eBook)
DOI 10.1007/978-1-4899-6675-9

To Glynis and Christa

Contents

Series Editor's Foreword xi

Preface xii

Section I 1

1 **General Introduction to Microelectronics** 3
   1.1 History of Microelectronics 3
   1.2 The Current and Future State of Microelectronics 9
   1.3 Production of an Electronic System 10
   1.4 Systems-on-a-chip 11
   1.5 Layout of Book 13

2 **Introduction to MOS (Metal–Oxide–Semiconductor) Devices and Logic** 14
   2.1 The MOS Transistor: Physical Behaviour 14
      2.1.1 The MOS Transistor Switch: Descriptive 16
      2.1.2 The MOS Transistor Switch: Device Equations 19
      2.1.3 MOS Transistor Types 22
   2.2 The Complementary MOS (CMOS) Inverter 24
      2.2.1 The CMOS Inverter: Logical Performance 24
      2.2.2 The CMOS Inverter: Speed Performance 26
   2.3 CMOS Logic 32
      2.3.1 CMOS Logic Gates: Logical Performance 32
      2.3.2 CMOS Logic Gates: Speed Performance 35
   2.4 nMOS Logic 35
      2.4.1 nMOS Inverter: Simple Analysis 36
      2.4.2 nMOS Logical Performance 39
   2.5 Synchronous Logic: An Introduction 41
   2.6 Summary 42

3 Fabrication of Silicon Integrated Circuits — 43
  3.1 Wafer Production — 44
  3.2 Production of Integrated Circuit Masks — 44
  3.3 Photolithography — 46
  3.4 Oxidation — 48
  3.5 Diffusion — 48
    3.5.1 Constant Source Diffusion — 50
    3.5.2 Limited Source Diffusion — 50
    3.5.3 Practical Diffusion Techniques — 52
    3.5.4 Typical Diffusion Apparatus — 52
    3.5.5 Lateral Under-Diffusion — 53
  3.6 Ion Implantation — 54
  3.7 Low Pressure Chemical Vapour Deposition (LPCVD) — 57
  3.8 Metallisation — 59
  3.9 Example of nMOS Process — 60

4 Design Rules — 63
  4.1 Contents of Design Rules — 63
    4.1.1 Geometric Design Rules — 63
    4.1.2 Electrical Design Rules — 66
    4.1.3 Mandatory Features — 66
  4.2 Process-Independent Geometric Design Rules — 66
    4.2.1 Use of Buried Contacts to Define Transistor Gate Lengths — 72
  4.3 Process-Independent Electrical Design Rules and Mandatory Features — 72
  4.4 Advantages and Disadvantages of Process-Independent Design Rules — 73
  4.5 Consequences of Breaking Design Rules — 74

5 Other Integrated Circuit Technologies — 75
  5.1 Silicon Bipolar Technologies — 75
    5.1.1 Resistor Transistor Logic (RTL) — 76
    5.1.2 Diode Transistor Logic (DTL) — 77
    5.1.3 Transistor Transistor Logic (TTL) — 78
    5.1.4 Emitter Coupled Logic (ECL) — 79
    5.1.5 Integrated Injection Logic (I^2L) — 82
  5.2 Silicon-on-Sapphire (SOS) — 82
  5.3 Gallium Arsenide (GaAs) — 83
  5.4 Comparison between Logic Families — 84

6 Integrated Circuits: from Concept to Silicon — 86
  6.1 Gate Array/Masterslice Design — 86
    6.1.1 Gate Arrays with Wiring Channels ('Streets and Houses') — 87
    6.1.2 Sea-of-Gates Array — 88

	6.1.3	Gate Arrays: Advantages and Disadvantages	89
6.2	Standard Cell Design	90
	6.2.1	Standard Cell Design: Advantages and Disadvantages	92
6.3	Full Custom (Handcrafted) ASIC Design	93
6.4	Programmable Logic Array (PLA) Design	95
6.5	Summary and Conclusions	98

7 Design Discipline and Computer Aided Design (CAD) and Test — 100

7.1	Hierarchical Design and Macros	102
	7.1.1 Top-down Design and Partitioning	103
7.2	Design Verification by Simulation	104
	7.2.1 High Level Simulation	105
	7.2.2 Gate Level Simulation	106
	7.2.3 Switch Level Simulation	108
	7.2.4 Device Level Simulation	110
7.3	Physical Design and Chip Layout	110
	7.3.1 Traditional Graphic Editors	111
	7.3.2 Symbolic Editors	111
	7.3.3 Automatic Routing and Placement	111
	7.3.4 Silicon Compilers	112
	7.3.5 Circuit Extraction and Re-Simulation	113
	7.3.6 Generation of Fabrication Data	113
7.4	Integrated Circuit Testing	113
	7.4.1 Test Pattern Generation	113
	7.4.2 Test Pattern Validation	115
	7.4.3 Design for Testability	116
	7.4.4 Application of Test Patterns	116

Section II — 117

8 The GATEWAY Gate Array Design Exercise — 119

8.1	Introduction	119
8.2	The GATEWAY Philosophy	119
8.3	The ED500 Gate Array	120
	8.3.1 The 'Hardwired' Inverter M3/M4	124
	8.3.2 Use of Transistors M1 and M2	125
	8.3.3 ED500 Gate Array Performance	129
8.4	The ED500C Gate Array	130
	8.4.1 The 'Hardwired' Inverter M3/M4	132
	8.4.2 Use of Transistors M1 and M2	133
	8.4.3 ED500C Gate Array Performance	135
8.5	ED500/ED500C Speed Performance Calculations	139
8.6	Design Methods, Discipline and the GATEWAY CAD	140

8.7 GATEWAY Project Assignment		141
8.7.1 Circuit Description and Details		141
8.7.2 Project Management		143
8.7.3 Costing of Project		145

Appendix: Inverter Output Rise Time   *146*

Bibliography   *148*

Index   *149*

Series Editor's Foreword

The rapid development of electronics and its engineering applications ensures that new topics are always competing for a place in university and polytechnic courses. But it is often difficult for lecturers to find suitable books for recommendation to students, particularly when a topic is covered by a short lecture module, or as an 'option'.

Macmillan New Electronics offers introductions to advanced topics. The level is generally that of second and subsequent years of undergraduate courses in electronic and electrical engineering, computer science and physics. Some of the authors will paint with a broad brush; others will concentrate on a narrower topic, and cover it in greater detail. But in all cases the titles in the Series will provide a sound basis for further reading of the specialist literature, and an up-to-date appreciation of practical applications and likely trends.

The level, scope and approach of the Series should also appeal to practising engineers and scientists encountering an area of electronics for the first time, or needing a rapid and authoritative update.

Paul A. Lynn

Preface

There are several excellent textbooks available that treat Integrated Circuit (IC) Design at a level appropriate to final year students, to practising designers or to postgraduates. Some of these advanced texts have an 'engineering' bias, and give a detailed treatment of the *technology* underlying integrated circuits. Others have a 'computer' science systems' leaning, concentrate on *methods* of IC design, and do not discuss the underlying technology beyond the barest fundamentals. These textbooks are all expensive, and their use is only justified in an advanced course, where they form one of the major (or even *the* major) text for the course.

Integrated circuit technology has now reached a stage of maturity where it forms an integral part of most undergraduate courses in Electrical Engineering, and indeed many Computer Science and Physics courses. It is therefore no longer appropriate to leave IC design as merely a 'final year option'. It is necessary to introduce the subject at a much earlier stage via a 'primer' course, leaving advanced material to a later course, which only enthusiasts need attend. There is at present no inexpensive text suitable for introducing pre-final year (or younger) undergraduates to IC Design, and it is irresponsible to recommend an advanced text for an introductory course. This book is intended to be a 'first' book of IC design that can stand alone as a text for non-specialists, or can lead to a more advanced course in a subsequent year of study.

The first chapter of the book puts the development of microelectronics in historical perspective, so that its meteoric pace may be appreciated. The focus of the book is on design, and the technological details are restricted to those essential to performing and understanding a simple IC design. However, the book is intended to be a true introduction, which would be incomplete without some underlying details. We have assumed only a knowledge of elementary Boolean algebra, of the barest details of electron/hole physics, and of simple linear circuit analysis (Ohm's Law!). The book is therefore self-contained for the reader who understands these fundamentals. From this basis, the physics of MOS transistor devices is discussed in chapter 2, in a simplified form appropriate to what we need to get out of it at this stage.

Chapters 3 and 4 take the same attitude to IC technology as chapter 2 took to device physics. Chapter 3 describes the processes by which MOS devices are fabricated. This treatment allows us, in chapter 4, to extract from the apparent alchemy of the fabrication process a set of specifications that allow a designer remote from the manufacturing site to prepare correct IC designs for fabrication. These are the 'design rules' for the process. Although this book deals in detail with only Metal-Oxide-Silicon (MOS) technology, other IC technologies have application areas which, although smaller than MOS, should not be disregarded. Chapter 5 catalogues the more important 'alternative' IC technologies, and highlights their strengths and weaknesses.

There are many ways of taking an IC specification or concept through the IC design process to produce a working chip. Each method has its advantages and shortcomings, and although section II of this book concentrates on only one of these methods, it is essential that the others be understood. Chapter 6 discusses the three main classes of IC design approach, and indicates the considerations of performance, cost and expediency that lead to a choice.

The design of a complex integrated circuit containing millions of transistors is a difficult task, which would be impossible without the correct use of Computer Aided Design (CAD) tools. The importance of these tools in the IC design process has become such that they have partially dictated the speed and direction of further development. Even an introductory text such as this must impart a feel for the range and scope of the CAD techniques for IC design. While chapter 7 makes no attempt to deal with the inner workings of CAD software, it does describe the major areas of IC design where CAD is necessary. In addition, discussion of IC testing is included in this chapter, as the field of test is so heavily reliant upon good supporting CAD. The need for discipline and care in IC design is acknowledged universally, and the fundamental techniques of hierarchical design and extensive simulation are ubiquitous. Chapter 7 explains how CAD tools encourage, aid, and occasionally enforce good design practice.

In learning any new skill, there is no substitute for practice. Integrated circuit design is no exception to this rule, and it is the aim of this book to equip its reader to perform a simple IC design. The first section of the book (chapters 1-7) constitute the background and technological detail essential to such an exercise. The second section comprises, in one chapter, a design 'manual' for a Gate Array design exercise ('GATEWAY') that has been used at the University of Edinburgh since 1982. The exercise is set in industrial terms, and includes timescales, costing details and other constraints such as would be encountered in an industrial setting. The project is designed to fit the academic year, and is used as part of the laboratory/course work at Edinburgh, in parallel with other activities. The assignment itself is deliberately straightforward and the circuit small, as we have found no benefit in the use of a large design. Indeed the effect of a large IC design is likely to be negative, as a clear view of the conceptual material is obscured by the sheer size of the task. It is our experience that the industrial setting, and the feeling of designing to a 'real' specification produces a positive reaction from students. Healthy competition grows to see who will do the 'best' design, and the learning potential is enormous. There is a set of software tools available to enhance the GATEWAY exercise, although it can be used without them. The software can be obtained from Unived Technologies Ltd, 16 Buccleuch Place, Edinburgh EH8 9LN.

In summary, this book should arm a pre-final year student with the knowledge to ease his passage into and through a more advanced IC Design course and textbook. In addition, it can stand alone to give a complete appreciation of the essentials of IC design to the non-specialist engineer or scientist.

The authors are grateful for the encouragement and advice of all their colleagues and of students who have performed the GATEWAY exercise. In particular, we wish to acknowledge a debt to Professor Mervyn Jack, the original architect of the GATEWAY exercise. The GATEWAY software was written with the help of Timothy Lees, and the microcomputer-based full custom graphic editor with the help of Frances Cranston. Tony Bramley drew the outline diagrams for the figures used throughout the book. We thank Professor John Mavor for his support as an advisor, and also in his capacity as Head of the Electrical Engineering Department at the University of Edinburgh. The collective expertise of the department, the friendship of the rest of its members and the overall environment have played no small part in the history of this book.

University of Edinburgh Alan F. Murray
Dept of Electrical Engineering H. Martin Reekie
The King's Buildings
Edinburgh, EH9 3JL

Section 1

1 General Introduction to Microelectronics

1.1 History of Microelectronics

We can hardly imagine an existence without the modern 'miracles' of computers and instantaneous audio and visual communications. The sophisticated control systems available to us as part of many rather mundane labour-saving devices such as automatic washing machines have also been totally absorbed into today's accepted way of life. All these things have been made possible only by the recent huge advances in electronics.

Electronics before the advent of semiconductors

The first major use of electronics was in radio. At that time radio transmitters were just large spark generators and produced signals which spread over a very large frequency spectrum. This limited the number of stations which could be 'on the air' simultaneously. J. A. Fleming's invention of the thermionic diode in 1904, together with Lee De Forest's subsequent invention of the *triode* two years later, revolutionised radio transmitter design. The triode was critically important because it was the first device capable of *amplifying* electronic signals. Using the triode (and its 'relatives', the screen grid valve and the pentode) radio transmitters which operated within a very narrow section of the frequency spectrum were designed. Cheap radio receivers were also produced. These early radios were not very good compared with modern sets because they used valves, the only available amplifiers, and so consumed a lot of electrical power. This was a major disadvantage because a domestic electricity supply was not widely available at the time and the batteries which powered these early radios frequently needed recharging. Later, when electricity was supplied directly to most homes, this problem was overcome and very large and complex radio receivers were developed. The market for these radios grew extremely rapidly and stimulated the further development of not only radio, but many other areas of electronics. For example, in the late 1920s and early 1930s, pioneering work was being carried out on television. The mechanical system, developed in 1927 by J. L. Baird, was quickly superseded by an all-electronic system using cathode ray tubes and the first public broadcast was made from Alexandra Palace, London, on 2nd November 1936.

The Second World War brought an end to the development of television but resulted in huge advances in other fields of electronics. For example, the first practical computer was constructed and was used to break the coded messages sent by the German ENIGMA system. The first signal processing systems, as we understand them today, were developed for use in war-time radar sets. The result was radars with greatly improved range and resolution.

All the electronic equipment constructed at this time had serious problems. It was unreliable and suffered from excessive size, weight, power consumption and cost. These problems could not be overcome because the only available amplifier was the valve.

The semiconductor age

The limitations imposed by valves on the size, weight and power consumption of electronic equipment were widely recognised and this led to several teams of researchers being set up to find alternative amplifiers. In 1947, in Bell Labs in the United States, the age of semiconductors finally arrived when W. H. Brattain and J. Bardeen pressed two fine gold probes into the surface of a germanium crystal and made the first transistor. The probes had to be positioned very carefully and very close together, which resulted in a rather delicate device with a poor gain and bandwidth performance but it did show that a semiconductor device could amplify electrical signals.

Almost immediately afterwards W. Shockley, also of Bell Labs, proposed the junction transistor, a device that had no wire contacts and could thus be made to work reliably. Both the first transistor developed and these new junction transistors had charge carriers of both polarities operating simultaneously and were therefore *bipolar* devices. The charge carriers were electrons and *holes*, new types of particles which could only be explained using quantum mechanics. Holes carry a positive charge and can be viewed for our purposes as positions in the semiconductor crystal lattice where electrons should be present but are not.

Theoretical studies showed that reliable transistors could only be constructed from extremely pure single crystals of semiconductor. In about 1950, Teal of Bell Labs grew single crystals of germanium (and later, silicon) that had less than one part in a billion of impurity atoms and this allowed the intentional introduction of controlled amounts of particular impurities known as *dopants*. These dopants were either *donor* or *acceptor* atoms and they were introduced at levels of about one dopant atom to every one hundred million or so atoms of the original semiconductor. A *donor* atom has one extra electron in its outer shell when compared with the original semiconductor and so, when it tries to fit into the semiconductor crystal lattice, it finds that it cannot find partners for all its outer shell electrons and has one left over. This electron is not strongly bound to its nucleus and is therefore available as a carrier of charge.

An *acceptor* atom has one electron less in its outer shell than the atoms in the original semiconductor, and so it feels that it is short of one electron and tries to catch any that are passing. If a single electron is passed from acceptor atom to acceptor atom, charge is being transported from place to place. When we are working with semiconductors it is convenient to think of a positive charge (the hole) moving around, rather than viewing this charge transportation as a movement of negative charge. The situation is analogous to that of a bubble moving in a water-filled tube. While we know that it is the water that is moving, we watch the movement of the bubble with more interest!

The first junction transistor was made from a sandwich of acceptor doped, donor doped, acceptor doped germanium semiconductor. In 1950 the first grown junction transistors were developed and the alloy junction process appeared the year after that.

Integrated circuits

In 1959, at a convention of the Institution of Radio Engineers, Kirby announced the 'solid circuit', which later came to be known as the *integrated circuit*. His idea, which was to revolutionise electronics, was simply that most of the critical elements of an electronic circuit could be made as part of a single piece of semiconductor material. It was not only possible to make transistors, but also diodes, resistors, capacitors and interconnecting wire on this single piece of semiconductor. Of all the basic electrical components only medium and large values of inductance could not be made as part of an integrated circuit.

The production of integrated circuits was only possible because techniques had already been developed to allow the fabrication of transistors on just one side of the silicon wafer, because batch processing was already commonplace and because methods had been developed to passify the silicon surface by depositing on it a layer of silicon dioxide. This illustrates the fact that in the field of integrated circuit production progress has almost always been made through a gradual 'coming together' of ideas rather than because of some great breakthrough.

Field effect transistors

Before the invention of the bipolar transistor, some research had been carried out into the 'field effect', that is the change in longitudinal conductivity of a semiconductor caused by a transverse electric field. In fact, it was during this research that the bipolar transistor was discovered. However, it was 1951 before the junction field effect transistor was proposed formally by William Shockley. Unfortunately, early attempts to fabricate these devices failed because the surface of the semiconductor (where mobile charge carriers were found) could not be made sufficiently stable. This problem was overcome as part of the development of the planar process for bipolar transistors, one step in which was the growth of a thin (≈ 0.1 μm) layer of silicon dioxide on the silicon to passify its surface (to protect it from contaminants).

In the construction of a Field Effect Transistor (FET) a thin layer of silicon dioxide was grown on top of a clean silicon surface and on this was placed a metallic electrode (the *gate*). It was found that a voltage applied between this gate and the bulk semiconductor induced charge carriers near the surface of the silicon. A *source* and a *drain* diffused region were then placed such that the gate electrode covered the area between them, and it was found that the current beween the source and the drain was controlled by the voltage on the gate. The first such Metal Oxide Semiconductor Field Effect Transistor (MOSFET) was demonstrated by Kahng and Atalla of Bell Labs in 1960. Unfortunately the characteristics of the first devices were not reproducible and it was not until about five years later that the source of these problems was found to be contaminants (in particular, sodium) and techniques were developed to eliminate them.

The techniques that allowed the integration of many bipolar transistors, together with their interconnections, on one semiconducting substrate were also applicable to the integration of MOSFETs, and therefore MOS integrated circuits were soon being produced. The major advantage of these circuits was that they had a higher packing density than bipolar circuits which implied that an MOS chip of a given area could perform more functions than a bipolar chip of the same area.

Because of the advantages of device miniaturisation, low power dissipation and high yield MOS integrated circuits now make up a major part of the integrated circuit market and this book will therefore emphasise MOS-related technology.

Milestones of integrated circuit progress

Some of the more important milestones in the development of integrated circuits are as follows

1951 – Discrete transistors available commercially.
1960 – Small Scale Integration (SSI), between 3 and 100 components per chip.
1966 – Medium Scale Integration (MSI), between 100 and 1000 components per chip.
1969 – Large Scale Integration (LSI), between 1000 and 10 000 components per chip.

1975 – Very Large Scale Integration (VLSI), more than 10 000 components per chip.
1985 – Wafer Scale Integration. Not yet commercially available but made in laboratories.

By the mid 1980s a VLSI device was usually considered to be a chip which contained at least 10^5 transistors.

It was noted by Moore in 1964 that the number of components on a chip had doubled every year since 1959, when the planar transistor was introduced. He predicted that this trend would continue until the early 1980s, when it would slow to a doubling every two years. This prediction has proved to be remarkably accurate, and is likely to remain fairly accurate, at least until the 1990s (see figure 1.1).

The computer oriented market

The first computers were based on valve circuitry and were huge, inefficient, slow and unreliable. Their fate was sealed in the late 1950s when they were superseded by the first transistorised machines, the IBM 7090/7094 series. This

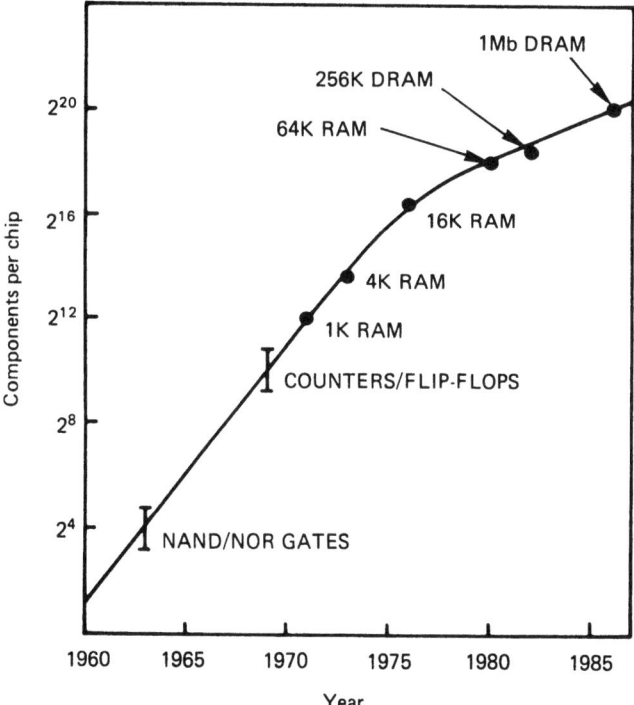

Figure 1.1 Moore's Law – the growth of complexity in integrated circuits

series dominated the market for many years until it was overtaken in 1965 by the third generation computers of the now famous IBM 360 series. Though these machines pioneered the use of integrated circuits in computers they still used magnetic core memories and it was only in 1970, when the 370 series was introduced, that IBM used semiconductor memories for the first time.

During the 1960s and early 1970s not all manufacturers followed the trend towards building faster and more powerful machines. Some concentrated on building small cheap computers that would be available to all, and this tendency was carried to its logical conclusion in the early 1970s with the introduction of the hand-held calculator. This small computer could perform the basic mathematical operations of addition, subtraction, multiplication and division, was sold for a few pounds, and had a figure of merit (expressed as performance per pound sterling) many orders of magnitude higher than the best computers of the early 1960s. Today the cost of producing a simple calculator is such that they are often given away free as part of advertising campaigns.

As the cost of computers has come down the cost of human labour has risen, and therefore it is becoming more cost-efficient to proceed along the path toward the electronic office and factory. The mini-computer is now becoming commonplace in even the smallest office where it is performing functions such as word-processing, data handling, accounting, financial management and more besides.

Computers in vehicles

The automotive industry is beginning to rely on the computer to help meet the standards of exhaust pollution, lower fuel consumption, higher performance and cheapness demanded by customers and governments throughout the world.

Confidence in the new technology has grown to such a point that manufacturers are willing to use microprocessors in that most critical of all automotive systems, the brakes. Antilock, anti-skid braking is becoming a major selling point for up-market cars, and it is to be hoped that it will not be too long before consumer demand and lower costs allow its inclusion in family cars.

Electronics in communications

The communications industry adopted solid-state electronics slowly at first, but now almost all equipment is transistor based. The new age of Information Technology requires the transfer of huge amounts of information from point to point and current resources are finding this difficult to handle. These problems are being solved by increasing use of microwave networks and optical fibre links, but because very high standards have been set for the transmission of both analogue and digital data across these networks they can be designed only by making extensive use of microelectronic devices.

Electronics in control systems

The use of sophisticated control systems has increased dramatically with the availability of modern electronics. Production processes and testing procedures have been automated, machine tools are computer controlled and 'user friendly' instrumentation which can be used by less highly-skilled operators has begun to appear. All these are possible only because of modern control engineering.

1.2 The Current and Future State of Microelectronics

The benefits of the increasing power and decreasing size of microelectronic circuits have been enormous – system sizes, weights and costs have been reduced by orders of magnitude and huge increases in system performance have been achieved. In fact the benefits of integration have become greater and greater as time has gone by, and the end is still not in sight. The size and complexity of the systems that can be built on silicon are now at such levels that their proliferation is currently limited more by the inability of humans and even the largest computers to comprehend fully the systems that they are able to construct.

Microelectronics is merging with the computer and communication industries to form the new field of 'Information Technology' (IT) or 'Information Engineering'. The two groups which initiated the study of Information Technology have been joined by a third: specialists in a new subject with the alarming title of 'Artificial Intelligence'. Those working in this new field aim to make computers 'think' in the same way as humans think (though perhaps computers should think a little more logically!). It is obviously desirable to build machines that can deal with the unexpected, can cope with concepts which are not precisely defined and can interact better with humans (by, for example, understanding the spoken word). Such machines will be made a reality by a combination of efforts from workers in IT disciplines.

Concurrent with research into Artificial Intelligence we can expect to see the development of machines with greater computing power, lower energy requirements and lower volumes which will allow the discoveries made by the Artificial Intelligence workers in the laboratory to be applied to every-day problems. If huge amounts of computing power become routinely available, many dreams which were previously firmly in the domain of the science fiction writer should be easily realisable.

Even from the examples given above it can be seen that the distinctions hitherto existing between a device, a circuit and a system have become blurred. The time of 'system-on-a-chip' is already with us; the only question for the future is how big the system on the chip will be.

1.3 Production of an Electronic System

There are at least four major techniques available for the construction of an electronic system.

Historically, the first to be employed was simply to build up the system from a number of discrete components. Not so very long ago the discrete components used would have included valves, but now these are used only in specialised applications where very high voltages and/or power are involved. A system constructed today would use discrete components only if medium voltages and/or powers were involved or if a near perfect linear system was required (for example, a very high performance hi-fi) and it would almost certainly be built using transistors. However, it is still possible to find tiny 'systems' which are constructed from discrete components simply because so little electronic 'work' has to be done. This book is concerned primarily with the design and manufacture of rather more complex systems than are possible (or desirable) using discrete components, and therefore we shall not consider this technique any further.

The second technique, which (justifiably) still finds favour in many systems, is to carry out the construction using commercially available Small Scale Integrated (SSI) circuits. This is a very popular method, primarily because it requires few special tools, is relatively simple to trouble-shoot and has low development costs. However, production costs can be very high as many integrated circuits must be bought and interconnected on complex printed circuit boards to construct the required system. There are also problems of design security — it is relatively easy for a competitor to unravel the secrets of a circuit constructed using SSI components. The technique's drawbacks can easily be tolerated, however, if the system complexity is low or only a limited production run is envisaged.

Obviously it would be beneficial if the large number of components mounted on an expensive printed circuit board could be replaced by something which was smaller and rather more versatile. Such a method does exist and it is the third major system design technique, the microprocessor.

The microprocessor is a very complex Very Large Scale Integrated (VLSI) circuit and its own development costs are astronomical. However, it can be programmed to perform many different tasks and thus a single type of microprocessor could be found in large numbers in many different systems varying from petrol pumps to washing machines. Because it can be used in so many applications, large numbers can be produced and its development costs can be spread over a very large number of chips. For this reason the cost of an advanced microprocessor can be in the region of a few pounds, despite the fact that a very large number of man-years of effort may have gone into its design.

Most microprocessors are not very good at communicating with the outside world and to do so successfully they need help. Usually they require to remember more program and more data than they can hold internally, so they also need help with their memory requirements. A microprocessor has no intrinsic idea of time, a concept which is vital to the operation of any system and to the microprocessor itself, so it clearly needs help there too. Help comes in the form of a number of supporting chips which will deal with the problems of data input/output, memory and clocking. Therefore, when we talk about a microprocessor we are actually talking about something which comprises a printed circuit board with a number of chips and a few discrete components mounted on it. The cost of this microprocessor *system* is much higher than that of the microprocessor *chip*, because of the cost of the board, the other components and the labour involved in assembling it. However, it is so versatile that this cost may still be acceptable.

The microprocessor is a small dedicated computer and, in the same way as a computer, it is useless unless it has been correctly programmed. Therefore the user will have to spend some considerable time and effort in developing a program which will cause it to perform the required function. He may also have to buy a microprocessor development system to allow him to carry out this work, and all of this adds to the system development cost.

All of the disadvantages mentioned above can be accepted because of the versatility of the microprocessor and the relative ease with which it can be reprogrammed to meet changing system needs. If a system has been constructed using SSI components and subsequently extra features are found to be desirable, or if the system must be altered to meet some slightly different requirements, it is often necessary to redesign it completely. This implies that the entire development cycle, with its attendant costs, must be repeated. However, if the system is based on a microprocessor it may be possible simply to alter the program it runs, which incurs a smaller redesign cost.

If a system is required to be extremely flexible and to be easy to update, there is little doubt that it would still be best implemented using a microprocessor. This is particularly true if the system is to be built in relatively small numbers or is to be on sale for a relatively short time. However, if large numbers of the system under consideration are to be made, it could be that the expense associated with the 'extra' bits around the microprocessor (the clocks, the memory, the interface with the outside world, the printed circuit board, etc.) would make it an uneconomic approach. If that is true then we should consider another approach which perhaps has higher development costs but has significantly lower production costs. Such an approach is the system-on-a-chip method which is introduced in the next section and is the main subject of this book. An integrated circuit designed to carry out a particular function is often called an Application-Specific Integrated Circuit (ASIC). This distinguishes it from a general-purpose microprocessor, but does not define the design method.

1.4 Systems-on-a-chip

There are at least three distinct approaches to putting as much of a system as possible on a single chip and they vary widely in design time, the level of design skill required, design costs, production costs and turnaround. However, the development costs for the chip are likely to be higher than those for any of the other techniques mentioned. The final chip can be hard to test and debug, and a variety of very specialised hardware and software tools are required to design, fabricate and test it. So why, then, does anyone bother to make an integrated circuit?

In commercial projects the search for a profit is the only good reason for an engineer to develop a system. Perhaps he wishes to design a system which is smaller and cheaper than that offered by his competitors so that his company will sell more of its products. Perhaps he wishes to design a system which will do something which has not been done before so that he will be able to sell it at a high price (and, hopefully, profit). Therefore, if a commercial engineer is to be persuaded to use integrated circuits he must be persuaded that their use will bring him economic benefits.

In fact, using integrated circuits can bring huge economic benefits, despite their apparently major disadvantages. The *production* cost of a single chip system is less than that of the same system constructed from small scale integrated circuits, generally by *several orders of magnitude*.

The reasons for using integrated circuits in military projects are usually rather different from those for using them in commercial projects. Military applications require equipment to be light (often portable), to have a low power consumption, to be extremely robust and reliable and to operate in some very harsh environments. Even the need to be able to destroy the equipment quickly in the event of its imminent capture may need to be considered by the design engineer. All these needs are best met using integrated circuits.

A system on a single integrated circuit has the following desirable features:

(1) The printed circuit board on which the system is mounted becomes far simpler and therefore cheaper. Fewer connections between chips are required and thus the costs of the labour and machines required to make the system are dramatically reduced.
(2) The replacement of many discrete components by an integrated circuit will dramatically reduce the number of interconnections in a system and, because interconnections are the part of a system which has the lowest reliability, this results in the overall reliability of the product being improved.
(3) The power consumption of the system will be greatly reduced, which may imply that smaller (cheaper) power supplies can be used.
(4) Since the power dissipation of the system has been reduced it will require less heat-sinking and may be cooled by natural air convention. This will reduce the system cost.
(5) The overall size of the system will have been reduced and savings can be made in the cost of its housing.
(6) A system constructed on a single integrated circuit will almost certainly operate faster than an equivalent system constructed from SSI components.

(7) The system design is much more secure if it is fabricated on a single chip. While chip designs can be (and are) copied, it takes a much greater amount of effort for a competitor to copy the design of a system-on-a-chip than a system-on-a-PCB.

It should be noted that the *overall* cost of the single chip system will be higher than that of the production cost of the chip, because development costs must also be considered, but still the overall cost advantages can be enormous if a medium or large production run is being considered.

1.5 Layout of book

This book is aimed at giving the reader the ability to perform correctly a simple Integrated Circuit Design Exercise known as GATEWAY, but before he can even begin such an exercise he must acquire a base of knowledge from which to work. The first seven chapters (section I) of this book should impart this knowledge, while section II of the book describes the design exercise in detail.

Chapter 2 covers the fundamentals of MOS transistor theory in some detail and considers both CMOS and nMOS technologies. It is necessary to understand this material (if not to remember all the equations!) before progressing further.

Many aspects of IC design are greatly affected by what is and what is not possible during device fabrication (for example, the minimum size of a transistor). These constraints are encapsulated in a set of *design rules* which must be obeyed by the IC designer. The MOS fabrication process is described in chapter 3 and a basic set of design rules is outlined in chapter 4.

The design study would be of little value if it was not seen in the correct context and thus caused the reader to be blinkered by the specific details of the GATEWAY exercise. Therefore it is essential that he should have some appreciation of the entire field of IC design. For this reason, chapters 5, 6 and 7 review the other technologies, design techniques and the Computer Aided Design (CAD) tools used in industry today. It may seem odd to include material on software in a book that is aimed firmly at the engineering aspects of microelectronics but it is essential that we do so, as CAD is so important to IC design that it cannot be omitted. Good CAD tools can make or break a designer, as they can remove many of the problems associated with the creation of a silicon chip which may consist of over a million transistors.

It is possible to make some headway into the GATEWAY exercise without reading all of the first seven chapters but, while it is accepted that there is no substitute for 'hands-on' experience to instil an understanding of the underlying concepts, the reader should beware of competely omitting the background material.

2 Introduction to MOS (Metal-Oxide-Semiconductor) Devices and Logic

This chapter presents a straightforward treatment of the principles of operation of MOS (Metal-Oxide-Silicon) transistors. While it is necessary to ensure that excessive simplicity (at the expense of correctness) is avoided, the essential aspects of MOS device behaviour should not be obscured by overly rigorous physics. The Bibliography at the end of the book suggests sources where the derivation and discussion of the full MOS device equations may be found. At present, however, to allow us to make a first attempt at MOS Integrated Circuit (IC) Design, we will concentrate on simplified forms of such equations as are essential to the discussion. In addition we shall use a 5V power supply. This is a typical value but is not immutable.

2.1 The MOS Transistor: Physical Behaviour

Before we attempt to use MOS transistors it is necessary to have some understanding of how they work. Figure 2.1 shows in cross-section an *n-channel enhancement* MOS transistor. In this type of MOS device, the starting material, or *substrate*, is silicon which has been doped with acceptors and is thus p-type. In a p-type semiconductor the majority charge carriers are positive holes.

The transistor has three terminals. The *source* and *drain* are regions of heavily doped n-type (called n^+) silicon and the section of substrate between them is called the *channel*. The third terminal, the *gate*, is formed by a conductive layer of polycrystalline silicon (often called 'polysilicon') which covers the channel and is separated from it by a thin layer of silicon dioxide (insulator). In earlier MOS technologies, this gate layer was made of metal, which led to the (now inappropriate but ubiquitous) Metal-Oxide-Semiconductor (MOS) label.

If no voltage is applied to the gate we have a sideways 'sandwich' of n-type silicon (drain), p-type silicon (channel) and n-type silicon (source). We therefore have two n-type to p-type junctions, each of which works as a diode, connected in series back-to-back. Because the two diodes are connected back-to-back, one of them has to be reverse-biased no matter whether the drain is positive or negative with respect to the source. No current can flow and the transistor is said to be 'OFF'.

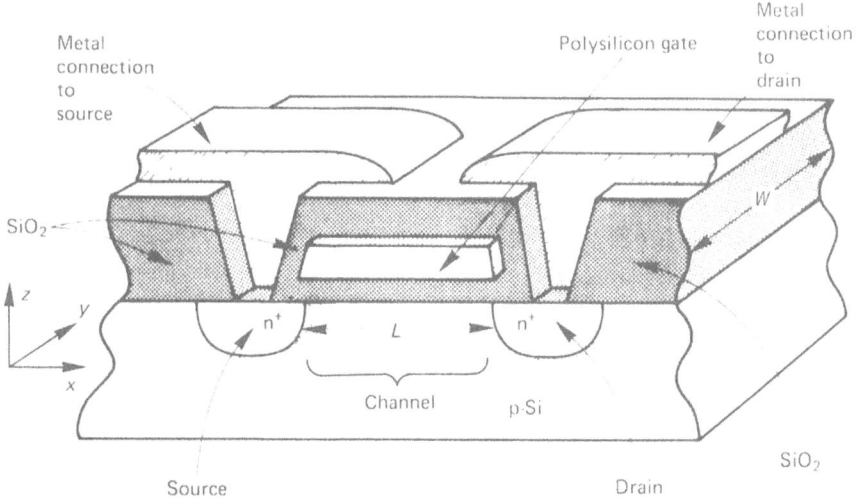

Figure 2.1 Cross-section of an n-channel enhancement Metal–Oxide–Silicon (MOS) transistor

Now in any semiconductor which is at a temperature above absolute zero ($0°K$) there is a finite statistical probability that an electron will break free from the atom to which it is bound, so producing an *electron-hole pair*. This is true regardless of the semiconductor doping and so even in the p-type substrate there will always be some free *electrons*. When the gate is made slightly positive with respect to the substrate these *minority* charge carriers are attracted to the channel. At the same time, *majority* charge carriers (holes) are repelled from the channel. Under these conditions the channel appears to be electrically rather less p-type than it was previously and it is said to be *depleted*. If the voltage on the gate is increased sufficiently there will be more electrons than holes in the channel region and the channel is said to be *inverted*. When inverted the channel appears *electrically* n-type, rather than p-type, and so the n-type ends of the back-to-back diodes (the source and drain regions) of the MOS transistor are shorted together, allowing current to flow. The transistor is said to be 'ON'.

This behaviour allows logic circuits to be constructed using MOS transistors, and we must therefore look at it in some more detail.

2.1.1 The MOS Transistor Switch: Descriptive

As stated in the previous section, if the voltage on the gate is zero, the MOS device can be compared to two back-to-back diodes. In the immediate vicinity of the p-n junction the channel is purged of charge carriers, and thus the source is isolated from the drain. This is shown in figure 2.2(a).

If the gate voltage, V_{gs}, is increased (becomes positive), *negative* minority charge carriers (electrons) are attracted from within the substrate into the channel. If V_{gs} is sufficiently positive, inversion may be reached (figure (2.2(b)). The critical value of V_{gs} at which the number of electrons at the surface just exceeds the number of holes, thus allowing conduction to start, is known as the *transistor threshold voltage*, V_t. If $V_{gs} > V_t$ the transistor is said to be 'ON'.

Now the amount of inversion at any particular point in the channel is dependent on the local conditions at that point and these may well be different in different parts of the channel. It is a reasonable first approximation to assume that conditions are equal along a line drawn across the device parallel to the source and drain edges (that is, normal to figure 2.2) but conditions will vary

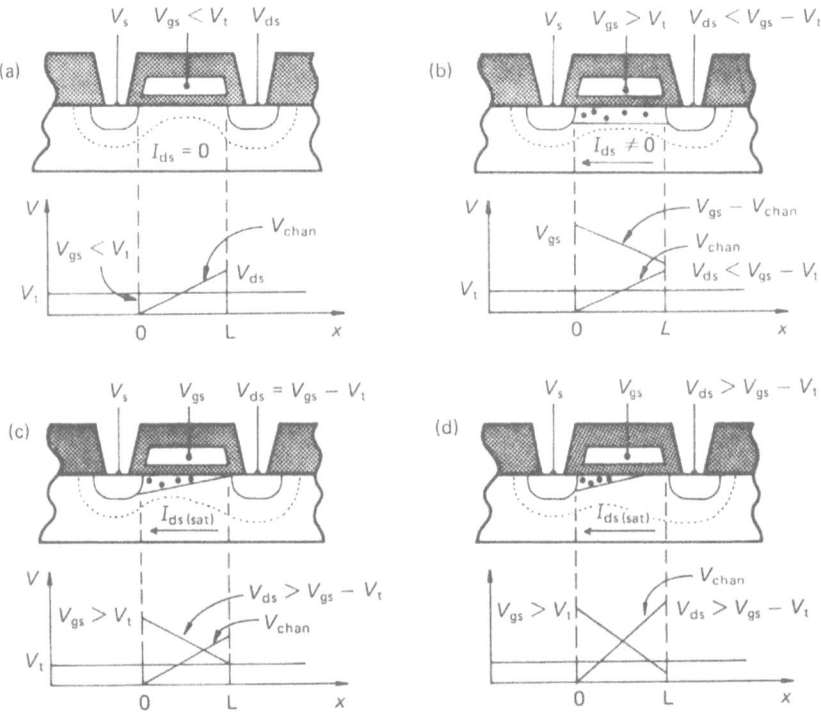

Figure 2.2 Cross-section of an n-channel enhancement MOS transistor in the 'OFF' (a), linear 'ON' (b), 'pinch-off' (c) and 'saturated' (d) operating regions

substantially along the length of the device. Obviously the degree of inversion in the channel at a given point (and thus the number of electrons present) is controlled by the voltage between gate and the channel *at that point*, which is given by $V_{gs} - V_{chan}$.

When V_{gs} is substantially greater than V_t and the drain-source voltage V_{ds} is small, the whole length of the channel will be inverted and current will flow. In fact, the channel acts (roughly) as a resistor, so the drain-source current I_{ds} can be approximated by Ohm's law

$$I_{ds} \approx V_{ds}/R \tag{2.1}$$

where R is the effective 'resistance' of the transistor.

This operating mode is known as the *linear* region. The 'resistance' of the channel is controlled by the gate voltage, as can be seen from figure 2.3. When the gate voltage on an MOS device is increased (and V_{ds} is small), the effective resistance $R(=\Delta V_d/\Delta I_{ds})$ decreases.

When the transistor is operating in the linear region the voltage on the channel at any particular point, x, along the length of the device is not difficult to find. The voltage on the channel right beside the source ($x = 0$) is clearly equal to 0 (we are assuming that the transistor source is grounded). Similarly the voltage on the channel right beside the drain ($x = L$) is equal to V_{ds}. The voltage on the channel will vary approximately linearly as x varies between 0 and L. Thus,

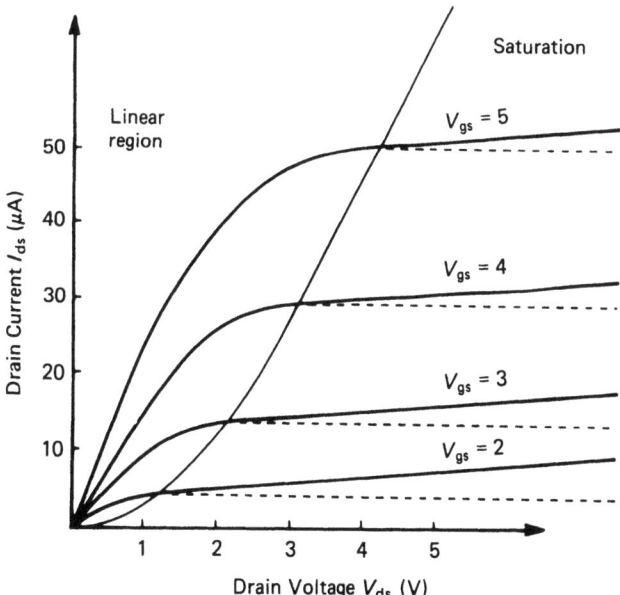

Figure 2.3 Characteristic curve of drain-source current against voltage for an n-channel enhancement MOS transistor for various gate voltages

for example, the voltage on the channel half way along the device will be $V_{ds}/2$.

Now let us consider local conditions on the channel right beside the drain region when V_{ds} becomes large. The voltage on the drain end of the channel will be V_{ds} and, of course, the voltage on the gate is V_{gs}. The voltage *between* the gate and the channel (which is the voltage which causes inversion of the channel) will be $V_{gs} - V_{ds}$ and for inversion to take place this must be greater than V_t. However, as V_{ds} is increased, the voltage between the gate and the channel at the drain end will reach V_t and so this section of the channel will *not* be inverted. This is shown in figure 2.2(c). Clearly, if V_{ds} is sufficiently large, the channel will not be inverted for some distance around the drain region, as shown in figure 2.2(d). Current therefore flows through a region of the channel which is not inverted. This happens because the channel near the source *is* inverted (assuming that $V_{gs} > V_t$) and the charge carriers (electrons in n-channel MOS) 'pick up speed' there and, if they are fast enough, they are injected into the non-inverted region as minority charge carriers. The high electric field in this region then sweeps them through to the drain.

The amount of non-inverted channel increases as V_{ds} increases. However, the speed built up by the electrons in the remaining inverted section of channel also increases (because of the increased electric field). We therefore have two opposing factors as V_{ds} increases; the electrons are asked to pass through a 'difficult' region (the non-inverted channel) but are given a higher initial velocity. There is a tendency for these two factors to balance out and so as V_{ds} is increased the *current* in the channel tends to remain constant. This mode of operation of the transistor is known as *saturation*.

If $V_{gs} > V_t$ and V_{ds} is increased from zero, the device is initially in the linear region. It enters the saturation region when V_{ds} has been increased to the point where the voltage between the gate and the drain end of the channel is just less than V_t (so that the channel at the drain is just failing to be inverted). The drain–source voltage at which the move from the linear to the saturation region takes place can be found by noting that at this point $V_{gs} - V_{ds} = V_t$. Reordering this equation gives the definition of the operating regions of the MOS transistor.

Linear region when $\qquad V_{ds} < (V_{gs} - V_t)$

Saturation region when $\quad V_{ds} > (V_{gs} - V_t)$

The first order theory and the equations presented in the following section predict that when the device is in saturation I_{ds} is constant as shown by the broken lines in figure 2.3. In reality, the shortening of the channel as V_{ds} increases (because of the increased width of the depletion region around the drain), coupled with the effects of electric fields in the substrate combine to give a slow but steady rise in I_{ds} (the solid lines in figure 2.3). In this book, however, and for all practical purposes, the current in saturation is taken to be constant.

2.1.2 The MOS Transistor Switch: Device Equations

Before proceeding to discuss the use of MOS transistors as logic elements, it is necessary that we fully understand the simplified forms of the device equations, and have at least *seen* them in unexpurgated form.

No attempt will be made to derive the full equations here, as this is treated elegantly in more advanced texts (see the Bibliography at the end of the book). The simplified forms will be made plausible, however, in so far as they describe switching.

We shall assume in this treatment that:

(1) Only drift current is considered.
(2) Carrier mobility is constant.
(3) Channel doping is uniform.
(4) Reverse leakage current is negligible.
(5) The transverse field E_z is much greater than the drain-source field E_x.

With these approximations/assumptions, the relationship between I_{ds} and V_{ds} is given by equation (2.2)

$$I_{ds} = \left[\frac{W}{L}\right] K' \left[V_{gs} - 2\Psi_B + \frac{Q_{ss}}{C_0} - \frac{V_{ds}}{2}\right] V_{ds} - \left[\frac{W}{L}\right] K' \frac{2(2\epsilon_s e N_a)^{\frac{1}{2}}}{3C_0} \left[(V_{ds} + 2\Psi_B - V_{bs})^{\frac{3}{2}} - (2\Psi_B - V_{bs})^{\frac{3}{2}}\right] \quad (2.2)$$

The parameters in equation (2.2) are:

- W: Width of transistor channel.
- L: Length of transistor channel.
- C_0: Capacitance/unit area of oxide under transistor gate.
- K': 'Process gain factor' = C_0 multiplied by electron mobility.
- Ψ_B: The different between the Fermi level in the substrate and the Fermi level for intrinsic silicon.
- Q_{ss}: Positive charge (/unit area) in surface states.
- ϵ_0: Permittivity of free space.
- ϵ_s: Relative permittivity of silicon.
- e: Electron charge.
- N_a: Concentration of acceptors in substrate.
- V_{bs}: Bias voltage on substrate, a voltage applied between the substrate and the ground connection. In an nMOS chip this voltage is negative.

This is a fearsome expression, but we should not allow its apparent complexity to dissuade us from proceeding. If this single equation is split into two in such a way that one part describes each of the *linear* and *saturation* regions, a simplification of equation (2.2) results which makes it much more usable.

For $V_{ds} < (V_{gs} - V_t)$ and $V_{gs} > V_t$ the transistor is *unsaturated* (in the linear region), and I_{ds} may be represented, to a good approximation, by

$$I_{ds} = \left[\frac{W}{L}\right] K' \left[(V_{gs} - V_t)V_{ds} - \frac{V_{ds}^2}{2}\right] \qquad (2.3)$$

where the threshold voltage V_t is given by

$$V_t = 2\Psi_B - \frac{Q_{ss}}{C_0} + \frac{2(\epsilon_0 \epsilon_s e N_a \Psi_B)^{\frac{1}{2}}}{C_0} +$$

$$+ \frac{(2\epsilon_s e N_a)^{\frac{1}{2}}}{C_0}\left[(2\Psi_B - V_s)^{\frac{1}{2}} - (2\Psi_B)^{\frac{1}{2}}\right] \qquad (2.4)$$

This threshold voltage is the same V_t as discussed in the preceding section, and is that voltage which, when applied to the transistor gate, results in the onset of channel inversion.

Equation (2.3) describes a curve with a maximum at $V_{ds} = (V_{gs} - V_t)$. This voltage, denoted $V_{ds}(\text{sat})$, marks the point where saturation begins. Equation (2.3) is inappropriate above this point because for larger values of V_{ds} the device is saturated and the current, I_{ds}, is constant. In other words, for all values of V_{ds} greater than $V_{ds}(\text{sat})$, where $V_{ds}(\text{sat})$ is given by

$$V_{ds}(\text{sat}) = V_{gs} - V_t \qquad (2.5)$$

I_{ds} is given by

$$I_{ds}(\text{sat}) = \left[\frac{W}{L}\right] K' \frac{(V_{gs} - V_t)^2}{2} \qquad (2.6)$$

In both equations (2.3) and (2.5) we require the value of the parameter V_t, and it would appear that it would have to be calculated from equation (2.4). In practice, a value for V_t (in volts!) is measured by those who fabricate the transistors and this value is supplied to the designer. The value should be regarded as a fixed process parameter (though the 'body effect', mentioned below, may also have to be considered).

For all practical purposes within the scope of this book, we may use equations (2.3), (2.5) and (2.6), disregarding the complexities of equations (2.2) and (2.4) with impunity.

There are only two 'second order' effects of which we should be aware. These are discussed below.

The body effect

In the preceding discussion on the simplified equations it was assumed that both the source terminal of the transistor and the transistor substrate were connected to ground. In fact, the substrate is often not connected to ground but to a negative voltage and, clearly, the transistor source need not always be connected to ground.

Let us consider a single n-channel transistor that has its source terminal grounded and discuss the changes in its behaviour when the substrate on which it is fabricated is connected to a negative bias voltage.

The transistor source and drain are regions of n-type silicon and the substrate is made from p-type silicon, so diodes are formed between each of the source and drain terminals and the substrate. These diodes are reverse-biased, so depletion regions will be formed at the junctions of the two types of semiconductor, preventing current flowing from the source or the drain into the substrate.

If the transistor is 'ON' a channel must be present under the transistor gate. In electrical terms, this channel looks like n-type semiconductor and, in the same way as for the n-type source and drain regions, current is prevented from flowing from it into the substrate by a depletion region. Therefore a depletion region surrounds the underside of the whole device, as shown in figure 2.2(c). This depletion region can be thought of as a very thin layer of insulating material 'protecting' the device from the negative voltage on the substrate.

What we have is a conducting region (the substrate) separated from the transistor channel by a thin insulator. This is a very similar situation to that of the normal transistor gate and so we can see that the substrate will act like a second gate connected to a negative supply. Now if we connect an n-channel transistor gate to a negative supply we will turn the transistor 'OFF' but, fortunately, the influence of this 'second gate' is not as strong as that of the normal gate and, instead of turning the transistor firmly 'OFF', it just *tends* to turn it off. Therefore, for us to turn the transistor 'ON' by a given amount while a negative voltage is connected to the substrate will require us to apply a little higher voltage to the normal gate than would have been required had the substrate been connected to ground. We can see that the major influence of the substrate back-bias is to *increase* the transistor threshold voltage.

In the above argument we assumed that the transistor source was grounded. This is not necessarily the case, but even so exactly the same argument applies. The voltage that is actually on the 'second gate' of the transistor is $V_{substrate} - V_{source}$ and is usually written V_{bs} (voltage bulk to source). This allows us to determine the degree to which the transistor threshold voltage is altered.

The increase in the transistor threshold voltage is given (approximately) by

$$V_t^* = V_t + \gamma(-V_{bs})^{\frac{1}{2}} \qquad (2.7)$$

Here V_t^* is the actual threshold value when body effect has been taken into account, V_t is the threshold value without body effect (when the transistor source and substrate are grounded) and γ is a process parameter measured by processing engineers. Typically γ is between 0.3 and 0.7 $V^{\frac{1}{2}}$.

I_{ds} in saturation

Figure 2.3 shows a non-zero slope in the I_{ds} against V_{ds} curves above $V_{ds} = V_{ds}(\text{sat})$. This effect is *not* implied by even the complete form of the equation for I_{ds} (2.2), as it is attributable to a combination of two further second order effects. The first is the shortening of the pinched-off channel above $V_{ds}(\text{sat})$. This induces a reduced channel resistance, and a subsequently increased current. The second effect is more subtle, and is the effect, through the substrate, of the strong electric field associated with the high drain voltage. This field acts as a positive 'back gate' on the channel, increasing the level of inversion, and thus reduces the channel resistance. Again, I_{ds} is increased above $I_{ds}(\text{sat})$.

In section 2.2, we shall see that in some cases, in particular when using CMOS, we can approximate even further from the simplified equations given above and treat the transistor as a resistive switch. Equations (2.3) and (2.6) are still required for more detailed calculations (and, indeed, equation (2.2) when the whole truth is necessary), but the 'resistive switch' approximation is good enough for many engineering purposes as, in general, we are performing 'worst case' design.

2.1.3 MOS Transistor Types

Thus far we have considered only n-channel enhancement devices, but there are other 'flavours' of MOS transistor which are essential to any form of MOS logic and these are introduced at this stage to set the scene for the design of MOS logic gates.

n-channel enhancement device

This (now familiar) device has the transfer characteristic (I_{ds} against V_{gs}) and symbol given in figure 2.4(a). The transfer characteristic shows clearly that the device operates as a 'switch', whose drain–source connection behaves as an *open circuit* ('OFF') for $V_{gs} < V_t$, and as a *medium-valued resistor* ('ON') for $V_{gs} > V_t$. The substrate voltage is shown as a terminal (bs) in the middle of the transistor symbol but this terminal is often omitted for clarity. The resistance of the 'ON' device is controlled by several factors, and is not constant. Equations (2.3) and (2.6) show that the most important factor influencing the 'ON' resistance is the device aspect ratio, W/L.

p-channel enhancement device

In this device, all the silicon types are interchanged (for example, the substrate is constructed of n-type silicon, while the drain and source are made from p^+-type silicon rather than n^+-type). Consequently, the charge carriers in the

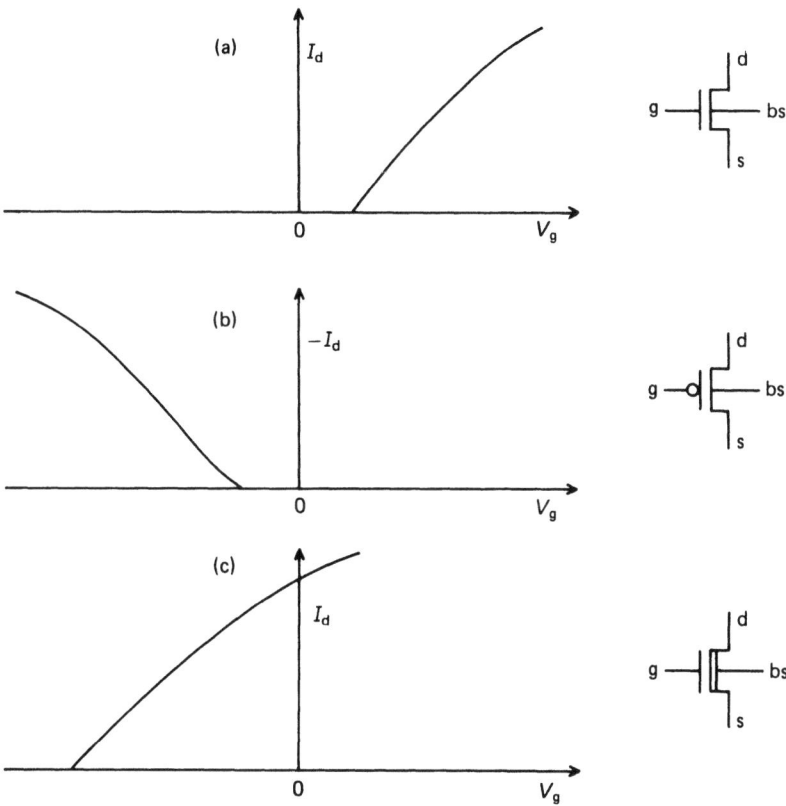

Figure 2.4 Turn-on/off characteristic curves and symbols for MOS transistors

channel are positive *holes*, and an inverting channel is formed when V_{gs} is *less* than V_t (which is negative). V_t for a p-type device is, of course, different from V_t for an n-type device, but in practice where n-type and p-type devices coexist, they will have about the same *absolute* value for V_t. More significantly, holes in silicon are about half as mobile as electrons and consequently p-type transistor widths (W in the equations) must be approximately doubled to give the same current-driving capability ('resistance') as their n-type counterparts (assuming constant transistor length, L). The p-channel transistor transfer characteristic is given in figure 2.4(b), along with its device symbol. Clearly, this device is analogous to the n-channel device, switching from open circuit or 'OFF' (when $V_{gs} > -|V_t|$) to conducting or 'ON' (when $V_{gs} < -|V_t|$).

Depletion devices

This form of MOS transistor differs from the types described above, in that it has a conducting channel *when the gate-source voltage is zero*. When a depletion transistor is processed it is implanted with donor atoms. This makes its channel slightly n-type even before any voltages are applied to its gate, and so to turn it 'OFF' a negative voltage must be applied to its gate to *deplete* its channel. Therefore the transistor threshold voltage (the voltage which must be applied to the gate to start to turn it 'ON') must be *negative*.

The transfer characteristic curve for an n-channel depletion device is shown in figure 2.4(c), along with the device symbol. Although p-channel depletion devices exist, they are so uncommon that they will not be considered further in this book.

The depletion device occurs most often with its gate and source connected. In this configuration, it is usually considered as a *resistor*, although it is actually not always of a constant value as it may be in or out of saturation, depending on the drain–source voltage.

We are now equipped with (roughly speaking) positive logic switches (n-type enhancement devices), negative logic switches (p-type enhancement devices) and resistors (depletion devices), all in an integrated form. In principle this enables the construction of logic circuits of arbitrary size and complexity. The next section describes how to abstract away from the detailed physical mechanisms of MOS devices to create CMOS circuits which will perform simple logic functions. This form of simplification is necessary if we are to avoid becoming confused by unnecessary detail at this stage in our IC design career.

2.2 The Complementary MOS (CMOS) Inverter

CMOS is currently the pre-eminent technology for new applications of VLSI (Very Large Scale Integration) circuits. This is because it couples high computational potential, immunity to environmental factors and conceptual elegance with low power consumption, efficient use of silicon and high speed. We shall therefore use CMOS as our first IC technology, and we will begin with the CMOS inverter. This logic element represents the fundamental circuit building block, and illustrates the principles behind the CMOS logic.

2.2.1 The CMOS Inverter: Logical Performance

Figure 2.5 shows the circuit and logic symbol for a CMOS inverter. It can be seen that it is made up of an 'upper' p-channel transistor and a 'lower' n-channel device. We must remember that a p-channel transistor works with the opposite polarities to an n-channel device and so, for example, its drain terminal is normally *negative* with respect to its source. Also to turn a p-channel transistor 'ON', V_{gs} must be *negative* and *less* than the transistor threshold voltage.

Now in the CMOS inverter shown in figure 2.5, the terminal of p1 which is connected to the 'OUTPUT' node must be more negative than the terminal

Introduction to MOS Devices and Logic

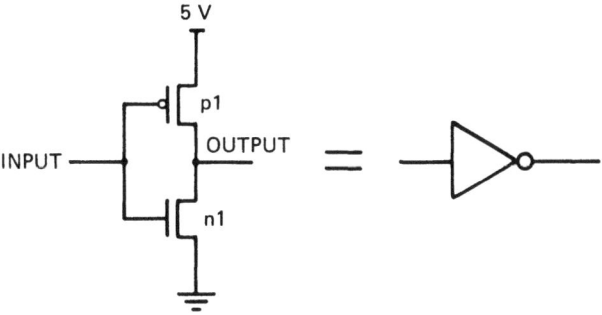

Figure 2.5 The Complementary MOS (CMOS) inverter

connected to the +5V power supply rail and so it must be the *drain*. Therefore, in this circuit transistors p1 and n1 are connected drain to drain.

If the INPUT terminal to the inverter is connected to 5V (logic '1') the lower, n-channel transistor will be turned hard 'ON'. Transistor p1 will be 'OFF' because it has 5V on both its gate and source and so the voltage *between* its gate and source, V_{gs}, is zero. The OUTPUT node is therefore connected to ground through n1 and isolated from the 5V power supply by the extremely high 'OFF' resistance of p1.

Conversely, if the INPUT terminal is connected to 0V (logic '0'), n1 will be 'OFF' isolating the OUTPUT node from ground. V_{gs} for p1 will be ($V_{gate} - V_{source}$) which is equal to $(0 - 5) = -5V$ and is *less than* the threshold voltage $(= -1V)$. p1 will therefore be hard 'ON', connecting the OUTPUT node to +5V.

Thus, OUTPUT = NOT(INPUT), and we have a logic inverter. To emphasise this operating mechanism, and to anticipate the construction of more complicated logic elements, we present the above discussion in tabular form as table 2.1.

Table 2.1 Operation of CMOS Inverter

INPUT	n1	p1	OUTPUT
0	off	on	1
1	on	off	0

Figure 2.6 shows how the OUTPUT voltage of a CMOS inverter is related to its INPUT voltage. Between the two extremes described in the preceding paragraph lies a region of transition, in which both n1 and p1 are 'ON'. Power is dissipated during this transitional phase, by the current flowing through the conducting devices (I_{ds} in both transistors). No power is consumed before the transition, however, as n1 is 'OFF' (open-circuited), and none is consumed after, as p1 is

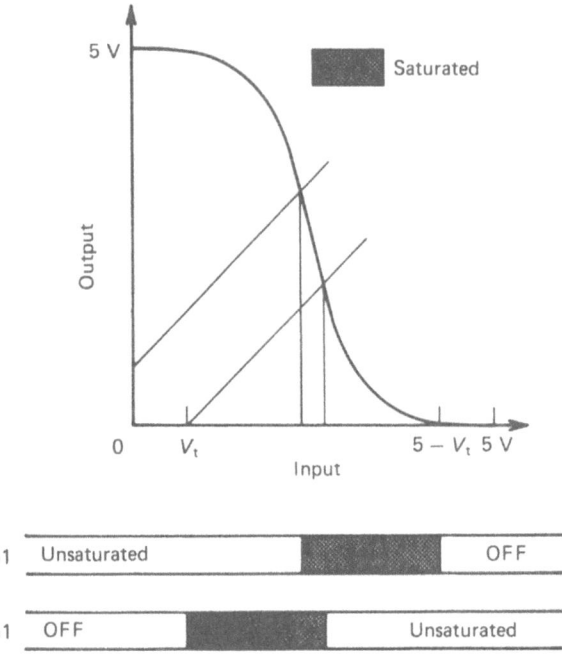

Figure 2.6 Input–output characteristic for a CMOS inverter

open-circuited. Because it is only during the instant of switching that power is drawn, more power will be consumed as device switching activity increases. In fact, the power consumed by a CMOS circuit is almost directly proportional to its frequency of operation.

For future reference, figure 2.6 also shows where p1 and n1 switch between their 'linear', 'saturated' and 'OFF' regions.

We can see from the above discussion that a CMOS inverter will indeed invert. To engineer a chip properly, however, we need to be able to design for adequate timing performance, as well as for logical correctness. In most applications an inverter would be useless if it took some seconds for its output to settle to the correct state.

2.2.2 The CMOS Inverter: Speed Performance

The CMOS inverter is shown again in figure 2.7, this time with a capacitance of C_i Farads attached to its output node. (Normally C_i would be measured in pico-Farads, rather than Farads!). It is the ability to charge and discharge this capacitance that determines the *performance* of an inverter. There are several contributions to this capacitance, but at present we shall consider only the two most important.

Introduction to MOS Devices and Logic

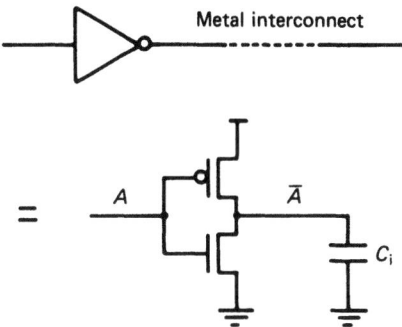

Figure 2.7 A CMOS inverter driving a load capacitance

Metal interconnect capacitance

After the transistor has been fabricated it is covered with a layer of silicon dioxide (insulator) to ensure that any conductive layers that run over the device do not cause any spurious short-circuits. However, it is obviously necessary to make some electrical connections to the transistor source and drain. From the schematic diagram of a transistor given in figure 2.1 it can be seen that holes have been cut through the insulating oxide to allow a conductor deposited on it to make contact with the underlying source and drain terminals of the transistor. These holes are called 'contact cuts'. During the fabrication of the device, metal is deposited over the entire surface of the wafer and it is then selectively removed, leaving strips of metal connecting different transistors together so that a complete circuit can be constructed. The metal strips on a chip are analogous to those on a printed circuit board. A strip of metal interconnect is separated from the substrate (which is at ground potential in AC terms) by the layer of insulating oxide and thus it forms one plate of a capacitor, the other side of which is grounded. This capacitance, C_i, is proportional to the area of the metal track.

Diffusion capacitance

In a standard IC every node in the circuit will be connected to a diffusion area; most probably to at least one transistor source or drain. This diffusion area is one end of a *reverse-biased* diode, and the other end is the substrate. This reverse-biased diode behaves as a capacitor because it has 'plates' (made from the diffused region and the substrate) separated by a dielectric (the depletion region formed because of the applied voltage). The value of the capacitor is set by three factors: the size of the diffusion area, the length of its perimeter and the applied voltage. The first of these is 'obvious'; it just gives the physical size of the capacitor plates and is thus directly proportional to the capacitance. The length of the perimeter is important because the bottom plate, the substrate, is very much larger than the top plate and thus fringing fields 'reach out' from beyond the edges of the top plate and tend to increase the effective plate area. Clearly this effect is dependent on perimeter length. The applied voltage, the last of the effects, varies the width of the depletion region, thus altering the distance between the 'plates' and varying the diode capacitance. This last effect is used in *vari-cap diodes* which are used as electrically variable capacitances.

The value of the capacitance of a diffusion area will therefore change with the voltages on the relevant circuit node. However, when analysing the circuit it is normal to find the largest possible value of this variable capacitor and then to replace it with a fixed capacitor of that value. The diffusion capacitance can be substantially greater than that of short lengths of metal interconnect.

Generally diffusion capacitance is undesirable, as it only adds to the stray capacitance in the circuit and reduces its speed of operation. If the reverse bias voltage on the diffusion/substrate junctions is *increased* the width of all the depletion regions at these junctions will increase and the junction capacitances will be *reduced*, allowing a higher maximum speed of operation for the circuit. In an nMOS IC this is achieved by connecting the substrate to a *negative* voltage. Note that this also has an effect on the transistor threshold voltages (see section 2.1.2 — The body effect).

Transistor gate capacitance

The output metal 'track' from a logic gate will, in general, lead to the input of another logic gate, but for simplicity we will assume that both the driving and driven gates are inverters. This is shown in figure 2.8, where substrate connections are shown explicitly. Let us now consider the performance of inverter (i). Its output is connected to the gates of the devices n2 and p2 in inverter (ii). Each of these transistor gates consists of a polysilicon conductor separated from its respective section of substrate by a layer of insulating oxide. Each transistor gate thus presents to the output of inverter (i) a capacitance which is proportional to the gate area. Although the substrate under the gate of p2 is connected to V_{dd} (+5 V, d = 'drain') rather than to V_{ss} (0 V, s = 'source') both C_n and C_p represent capacitances that must be charged or discharged.

The total resultant capacitance connected to the output node of inverter (i) can therefore be regarded as the interconnect capacitance, C_i, plus the diffusion capacitance, C_d, in parallel with the gate capacitances C_n and C_p. This total capacitance, C, which must be charged and discharged through p1 and n1, is shown in the lower half of figure 2.8.

The true nature of logic switching in a CMOS inverter can now be discussed.

When the input to a CMOS inverter which is driving a capacitive load is as shown in the upper trace of figure 2.9, the inverter output voltage will vary as shown in the lower trace.

Introduction to MOS Devices and Logic   29

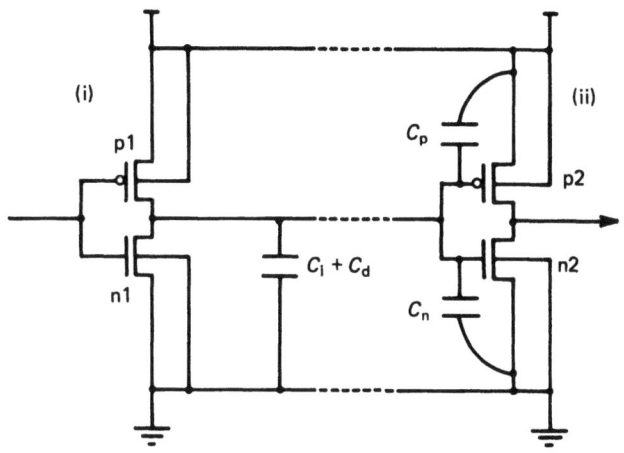

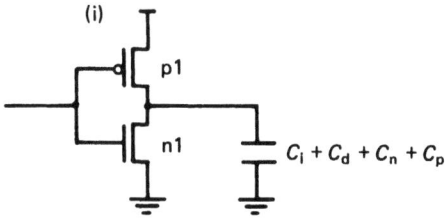

Figure 2.8 Cascaded CMOS inverters

During the non-switching static logic '1' and '0' states, $I_{ds} = 0$, and the capacitance C is irrelevant. However, while the output node is switching from logic '1' to logic '0' the capacitor C is being discharged, and while it is switching from logic '0' to logic '1' the capacitor is being charged.

During either of these operations a charge of $(V * C) = (5 * C)$ Coulombs is being transferred to or from capacitor C. Referring again to figure 2.9, the operation of the inverter may be detailed as follows:

$t < t_1$ The output is at a static logic '1', so $Q = 5C$ Coulombs is stored on C

$t_1 < t < t_2$ The inverter input is between V_t and $5 - V_t$, transistors n1 and p1 are both 'ON', and the output of the inverter begins to fall from 5V.

$t_2 < t < t_3$ Transistor p1 is 'OFF', the inverter input rises to 5V and C discharges exponentially through n1. After a time which depends on the value of C and the equivalent 'ON resistance' of n1, the charge on C will be almost zero.

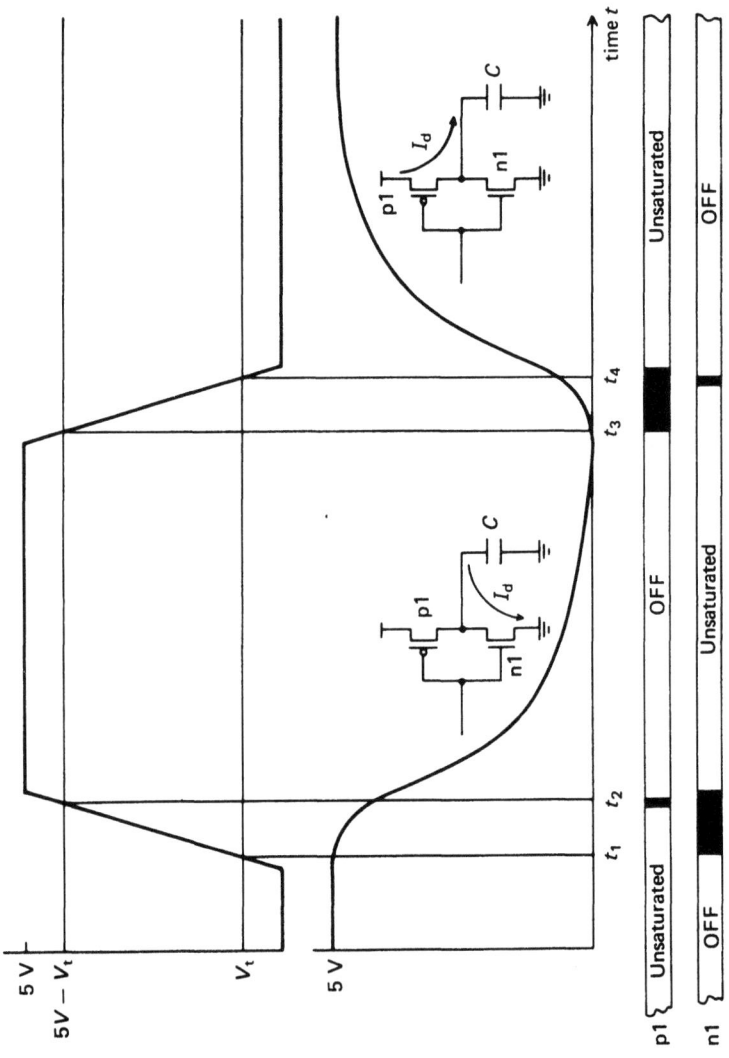

Figure 2.9 Input and output waveforms for a CMOS inverter driving a load capacitance

$t_3 < t < t_4$ The inverter input is between V_t and $5 - V_t$, transistors n1 and p1 are both 'ON', and the output of the inverter begins to rise.

$t > t_4$ The inverter input is 0V, and C is charged to 5V through p1. After a time which will depend on the value of C and the equivalent 'ON resistance' of p1, C will be storing a charge of $5C$ Coulombs.

The speed of operation of this CMOS inverter is dictated by the speed of charging or discharging the capacitance on the output of the inverter. As we have already said, this speed is dictated by two factors:

(1) The capacitance C connected to the inverter output.
(2) The equivalent 'ON resistances' of the transistors n1 and p1.

The capacitance C depends on the length of interconnect driven by the inverter and the number of other logic gates it drives (its *fanout*) and it is therefore straightforward to calculate. The 'ON resistances' of the transistors deserve closer inspection, however, as it is here that simplifications may be made.

The operating mode of the two transistors is shown below the timing curve in figure 2.9. These can be verified using the definitions in section 2.1. Clearly, during most of the switching operation, either transistor n1 is 'OFF' and transistor p1 is unsaturated, or vice versa. When this is the case, a constant capacitance C is being charged or discharged through the appropriate transistor equivalent 'ON resistance'.

Let us assume, for the moment, that transistors n1 and p1 have the same 'ON resistance', R. This resistance is a function of the aspect ratio of the device in question and can therefore be set to any desired value by modifying the transistor width and length. The time constant for the charge or discharge of the capacitor C will be $\tau = RC$. Careful use of the parameter, τ, allows an assessment to be made of the operating speed of *all* logic circuitry, provided a slight overestimate of τ is made to allow for the errors in this simple model.

As we conclude this impressive simplification of a physically complicated set of circumstances, it is worth sounding a note of caution in a summary.

CMOS inverter performance: summary

A CMOS inverter driving a load capacitance C switches within a time related to $\tau = RC$, where R is a function of the transistor geometry and a combination of process factors. We are therefore treating enhancement transistors as *RESISTIVE SWITCHES*. Furthermore, we are regarding n-type enhancement devices as switches that are 'ON' when the gate is at a logic '1', and p-type devices as 'ON' when the gate is at a logic '0'. This form of calculation is only safe when conservative estimates of R and C are taken, and should *not* be relied upon where speed performance is critical. Under those circumstances, full analogue simu-

2.3 CMOS Logic

Now that we are equipped with positive-logic and negative-logic switches, we can build more complicated logic elements.

2.3.1 CMOS Logic Gates: Logical Performance

Standard CMOS logic gates are simply extensions of the inverter, with multiple paths from the output node to V_{dd} or V_{ss}. Two simple examples are shown in figure 2.10, each driving an output capacitance. In the NOR gate shown, the output will be pulled down to a logic '0' if *either* n1 *OR* n2 (or both) is 'ON'.

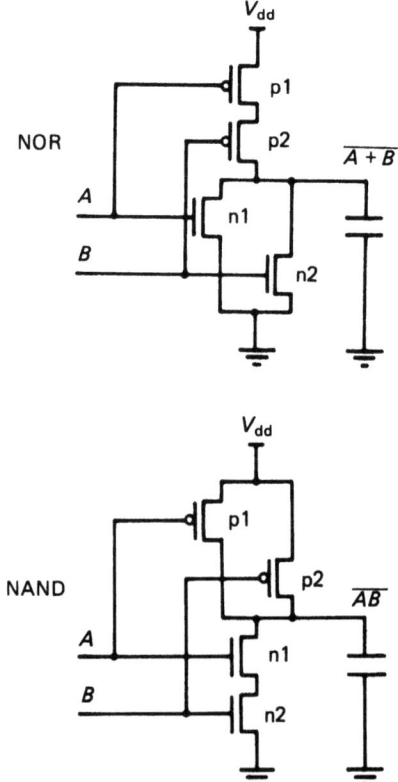

Figure 2.10 CMOS NOR and NAND gates

Under those circumstances, *either A OR B* (or both) is at a logic '1', and consequently *either* p1 *OR* p2 (or both) is 'OFF'. There is therefore no connection between the circuit output and V_{dd} (logic '1'). If *both A* and *B* are at a logic '0', *both* n1 and n2 are 'OFF', and *both* p1 and p2 are 'ON'. Only with this set of inputs *A* and *B* will the output be a logic '1'. This verbose description of the functionality of the NOR gate is summed up in table 2.2.

Table 2.2 Operation of CMOS NOR Gate

A	B	n1	n2	p1	p2	OUTPUT
0	0	off	off	on	on	1
0	1	off	on	on	off	0
1	0	on	off	off	on	0
1	1	on	on	off	off	0

In normal CMOS gates, the logic function carried out by the p-channel devices is always the *logical dual* of the logic function carried out by the n-channel devices. For the NOR gate discussed above the n-channel function is (*A* OR *B*), carried out by transistors n1 and n2 in *parallel*. The p-channel function is (*A* AND *B*), carried out by transistors p1 and p2 in *series*. These two transistor networks are referred to as the 'pull-down' and the 'pull-up' logic 'trees' respectively. The symmetry inherent in this design style leads to a conceptual simplicity. The need to build two logic trees of transistors for a single logical function has, however, led to a proliferation of 'dynamic' CMOS design styles. These avoid the problem of redundancy, but are subject to other problems, in particular those of timing. In any event, they are outwith the scope of this book.

Also shown in figure 2.10 is a CMOS NAND gate. Once again the pull-down (AND) function is the dual of the pull-up (OR) function, and the operating scheme is as given in table 2.3.

Table 2.3 Operation of CMOS NAND Gate

A	B	n1	n2	p1	p2	OUTPUT
0	0	off	off	on	on	1
0	1	off	on	on	off	1
1	0	on	off	off	on	1
1	1	on	on	off	off	1

It is often necessary to perform logical functions more complex than NOR and NAND, and the CMOS design style described above copes naturally with this requirement. Figure 2.11 shows a CMOS exclusive-OR (XOR) gate, composed of *two* distinct CMOS logic elements. The first gate is a simple two-input NOR, but the second (made from n1-3 and p1-3) performs the non-primitive logic function

$$\text{OUTPUT} = \overline{AB + C}$$

When $C = \overline{A + B}$, the output of the first gate, the output of this second gate is

$$\text{OUTPUT} = \overline{AB + \overline{A + B}} = A\bar{B} + \bar{A}B$$

This is the exclusive-OR function, as can be verified from table 2.4.

Table 2.4 Operation of CMOS XOR Gate (figure 2.11)

A	B	n1	n2	n3	p1	p2	p3	OUTPUT
0	0	off	off	on	on	on	off	0
0	1	off	on	off	on	off	on	1
1	0	on	off	off	off	on	on	1
1	1	on	on	off	off	off	on	0

The important principle illustrated here is that arbitrarily complex logic functions can be built into the pull-up and pull-down logic trees of a CMOS gate. The example of the second gate in figure 2.11 allows us to build the exclusive-OR function $(A\bar{B} + \bar{B}A)$ directly without recourse to explicit inversion of A and B.

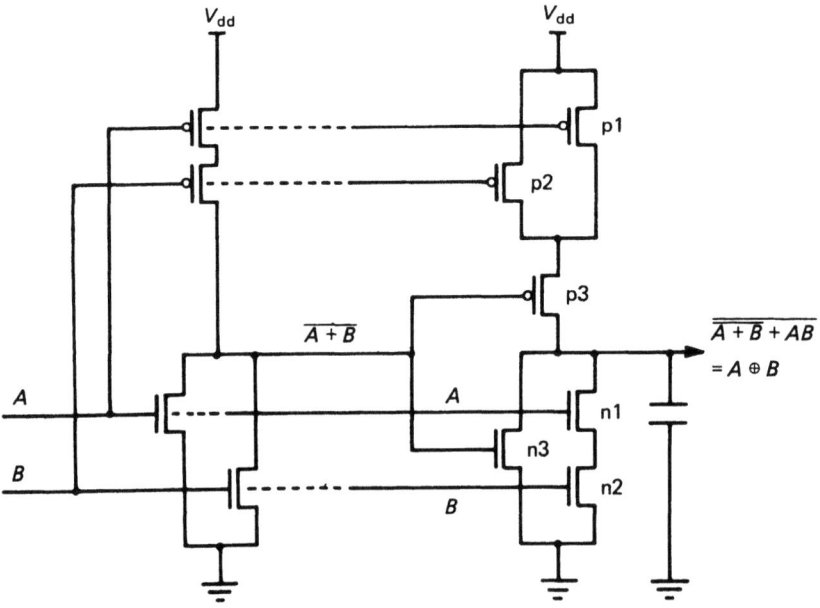

Figure 2.11 A CMOS Exclusive-OR (XOR) gate

The 'depth' of logic that can be carried out in this way is only restricted by the fact that greater logical 'depth' implies a lower speed of circuit operation. We can use the same simplification used in the examination of the inverter to estimate the speed performance of all CMOS logic gates.

2.3.2 CMOS Logic Gates: Speed Performance

If we visualise CMOS logic gates as networks of resistive switches, the delay owing to any CMOS gate can be estimated. With the CMOS inverter, switching meant charging or discharging the output capacitance through a single transistor of resistance R. In a logic gate, the resistance for charging/discharging will be different for different inputs to the logic gate. What we must do is to choose the 'worst case'. In the example of the NOR gate of figure 2.10 the worst case is either

(1) Charging the output capacitance through the series resistance of p1 and p2

or

(2) Discharging the output capacitance through *either* n1 or n2.

Of these two, the charging-up is the worst, as the time constant will be $2RC$ (R for p1 plus R for p2), while the time for discharge will be RC. *All* logic gates can be assessed in this way, provided care is taken to identify the worst case accurately. Estimates can be made of each of the series of delays that combine to produce the total delay through a series of CMOS logic gates. This total delay is crucial, as it restricts the number of calculations/operations per second that the logic can perform.

Designing IC logic which works correctly, in a logical sense, is easy. However, good IC engineering demands a careful tailoring of the speed of the circuitry to its intended speed of operation. This requires careful calculations, as described, of the speed of critical sections. In section II of this book which describes a particular IC design exercise, we give further details and a specimen calculation for the IC technology used.

2.4 nMOS Logic

The use of nMOS logic pre-dates that of CMOS. Although CMOS is fast becoming the 'mainstream' IC technology, nMOS is still popular, particularly when cost is a prime consideration. In an nMOS IC process, the transistors available are n-channel enhancement and n-channel depletion devices. The characteristics of these devices have already been given in section 2.1.3.

2.4.1 nMOS Inverter: Simple Analysis

The designer of an integrated circuit has available a number of fixed process parameters which he is expected to use to produce a geometric layout of the required IC. Because he is creating a *geometric layout*, one of his major interests is how to calculate the widths and lengths of the transistors that perform the logic operations. In a CMOS design the consequence of designing transistors which are of slightly incorrect widths or lengths will be that the speed of operation of the circuit will not be as expected. In an nMOS design the correct design of these transistor parameters is more critical since, if they are incorrect, it is possible that the circuit will not work at all. We must therefore investigate nMOS circuits in some detail.

Consider the simple inverter shown in figure 2.12. Once any load capacitance has been charged or discharged there will be no DC current flowing either to or from the output node. Thus the (steady state) drain-source current in the enhancement transistor, e1, must be equal to the drain-source current in the depletion device, d1, irrespective of the input to the circuit.

If the input to the inverter is a logic '0', I_{ds} will be zero for both transistors (transistor e1 is 'OFF'), and V_{out} will be 5V irrespective of the geometric sizes of the two devices. The situation becomes more complicated, however, if the enhancement device e1, if 'ON', when we would expect the inverter output to be a logic '0'. In this case current flows from V_{dd} through d1 and e1 to ground. Consequently, nMOS circuits *do* consume DC (static) power. This is one of their main disadvantages with respect to CMOS.

The output of the inverter will probably be connected to the input of another logic gate. If this output is to represent a logic '0' accurately, its voltage must be *less* than that required to turn 'ON' the next gate's pull-down transistor. Thus V_{out} must be less than V_{te}, the threshold voltage for an enhancement transistor.

Usually process parameters are adjusted in such a way that $V_{te} = 0.2 V_{dd}$ and so we have to ensure that V_{out} is less than this value. If V_{out} was exactly $0.2 V_{dd}$, the enhancement transistor of the next gate would start to turn 'ON' and the design would almost certainly fail. We have to be *certain* that V_{out} is

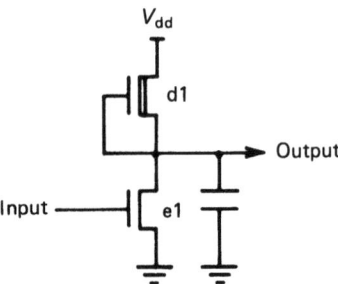

Figure 2.12 An n-channel MOS (nMOS) inverter driving a load capacitance

Introduction to MOS Devices and Logic

less than $0.2V_{dd}$ and the only way to do this is to build in a substantial 'margin for error' by designing the circuit so that V_{out} equals $0.1V_{dd}$. This safety margin of $(0.2V_{dd} - 0.1V_{dd}) = 0.1V_{dd}$ is called the *noise margin*.

In the example of figure 2.12 we have $V_{dd} = 5V$, so $V_{te} = 1V$ and we should design the circuit so its logic '0' output is 0.5V.

Here are some typical process parameters. These parameters will be used in this section and in the appendix, where the equations presented in the latter part of this section are derived. Radical departures from these values may require the results presented to be modified.

V_{te} (enhancement transistor threshold) = $0.2V_{dd}$
V_{td} (depletion transistor threshold) = $-0.8V_{dd}$
Noise margin = $0.1V_{dd}$

Let

$$\beta_e = \left[\frac{W_e}{L_e}\right] K'_e \text{ and } \beta_d = \left[\frac{W_d}{L_d}\right] K'_d$$

where parameters with the subscripts 'e' and 'd' refer to the enhancement device, e1, and the depletion device, d1, respectively.

We are considering the case where the input to the inverter is a logic '1' and so the output should be a logic '0' with allowance made for a suitable noise margin. Thus transistor e1 has V_{dd} on its gate, $0.1V_{dd}$ on its drain (the gate output) and has its source connected to ground. Therefore $V_{ds} = 0.1V_{dd} < (0.8V_{dd}) = (V_{gs} - V_{te})$ so transistor e1 is *unsaturated*. We therefore use equation (2.3) to calculate its drain current, I_{ds}. This gives

$$I_{ds} = \beta_e \left[(V_{dd} - 0.2V_{dd})0.1V_{dd} - \frac{(0.1V_{dd})^2}{2}\right] \quad (2.8)$$

Transistor d1 is a depletion device which has $0.1V_{dd}$ on its gate and its source and has V_{dd} on its drain. Therefore $V_{ds} = V_{dd} - 0.1V_{dd} = 0.9V_{dd} > 0.8V_{dd} = (V_{gs} - V_{td})$ so transistor d1 is *saturated* and we should use equation (2.6) to calculate its drain current, I_{ds}.
This gives

$$I_{ds} = \frac{\beta_d}{2} * (0 - (-0.8 * V_{dd}))^2 \quad (2.9)$$

The current, I_{ds}, in both transistors is equal, so from equations (2.8) and (2.9) we obtain

$$\beta_e * 0.075 * V_{dd}^2 = \beta_d * 0.32 * V_{dd}^2 \quad (2.10)$$

From equation (2.10) we see that

$$\frac{\beta_e}{\beta_d} = \frac{(W_e/L_e)K'_e}{(W_d/L_d)K'_d} = 4.27 \quad (2.11)$$

Integrated Circuit Design

Now the actual value for K'_d is usually a little less than the value for K'_e, and also the inaccuracies caused by using only first order equations have not been taken into account. Therefore a more practical version of equation (2.11), written in terms of actual transistor lengths and widths, would be

$$\frac{W_e}{L_e} = 4 * \frac{W_d}{L_d} \tag{2.12}$$

Using this fairly simple technique the IC designer can easily decide the *relative* sizes of his pull-up and pull-down devices (transistors d1 and e1, respectively).

Any load capacitance connected to the inverter output must be charged through d1 and discharged through e1. Now, because the aspect ratio, (W_d/L_d), of d1 is approximately one-quarter of that of e1, its current-driving capability is much less than that of e1. In other words, the equivalent resistance of d1 is much greater than the equivalent resistance of e1. The inverter rise time (the time taken for the inverter output to rise from logic '0' to logic '1' while charging C) is therefore much longer than the fall time and is the factor which limits the speed of operation of the circuit. This disparity between the charge and discharge (rise and fall) times is an unfortunate but unavoidable consequence of this *RATIOED* design style.

As the speed of operation of the circuit is defined by the current-driving capability of the pull-up transistor together with the circuit load capacitance, it is worth examining how these two components produce the circuit rise time. A detailed discussion is given in the appendix, but an outline of the argument is presented here.

Figure 2.13 shows how the output of the nMOS inverter of figure 2.12 rises from its logic '0' level of $0.1V_{dd}$ to its logic '1' level of V_{dd}. The input to the inverter would be 5 V for $(t<0)$ and zero volts for $(t>0)$ and so we can imagine that the pull-down device, e1, has 'disappeared' for $(t>0)$. Furthermore, we will ignore body effect and assume that the pull-up transistor has been fabricated on a 'normal' process so that its threshold voltage, V_{td}, will be $-0.8V_{dd}$.

The graph shows that the inverter output rises *asymptotically* towards V_{dd}. Therefore, to obtain a finite value for the rise time, we will say that the inverter output has risen when it reaches $0.9V_{dd}$.

A practical value for the time taken for the inverter output to reach $0.9V_{dd}$ is approximately

$$t_{rise} \approx \frac{7*C}{(\beta_d * V_{dd})} \tag{2.13}$$

Details of the mathematics behind the production of this equation are given in the appendix.

If $K'_d = 25$ $\mu A/V^2$, $V_{dd} = 5V$ and C is in pico-Farads then

$$t_{rise} \approx \left[\frac{56*C}{W_d/L_d}\right] \text{ nano-seconds} \tag{2.14}$$

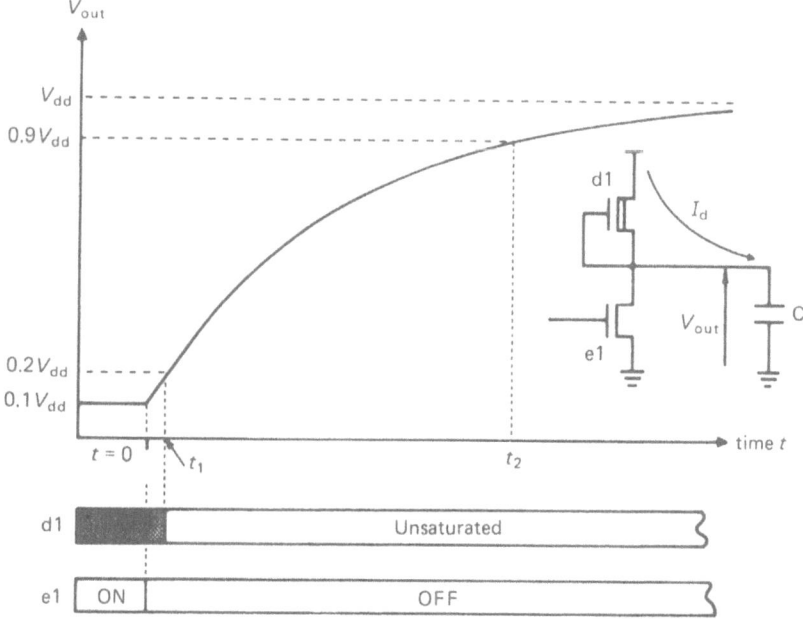

Figure 2.13 Rising output waveform for an nMOS inverter driving a load capacitance

The full analysis of the inverter rise time, taking into account second order effects, is complicated and is of little practical use, as the process parameters supplied to the designer all have large tolerances, which implies that any calculation to the last decimal place is pointless.

Estimates of nMOS logic gate rise times can be found by using one of the widely available circuit simulation packages such as SPICE. However, it is common practice to design nMOS circuits using equations (2.12) and (2.13). These allow us to calculate actual values for W_e, L_e, W_d and L_d if we note that it is normal practice to design every transistor so that *either* its length, L, or its width, W, is the minimum allowed by the rules of the particular process on which the design will be fabricated. We should also note that, because of *lateral under-diffusion* (see section 3.5.5), the *actual* length of a transistor is less than the *drawn* length. Further information on the magnitude of this effect on a particular process will be found in the process rules.

2.4.2 nMOS Logical Performance

Logic gates are constructed similarly to CMOS logic gates, except that only a pull-down logic tree is necessary. The nMOS NOR, NAD and XOR gates in figure 2.14 exemplify this. We shall discuss the XOR, which is composed of two distinct nMOS logic gates. If either e1 or e2 (or both) is 'ON', the output

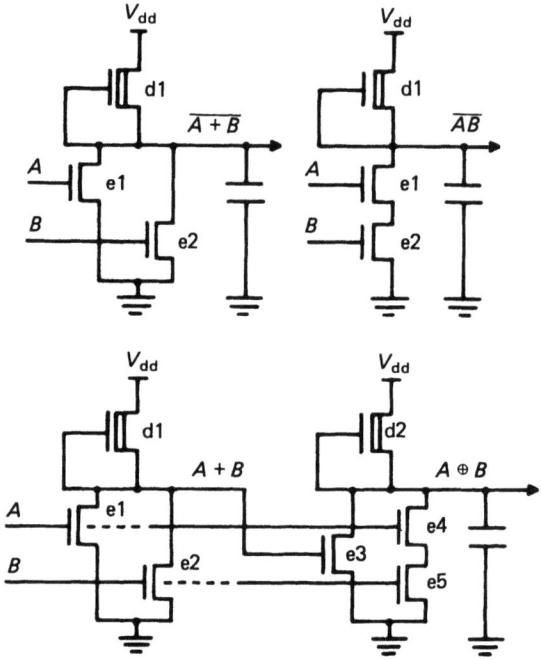

Figure 2.14 nMOS NOR, NAND and XOR logic gates

of the first (NOR) half of the XOR gate is logic '0'. If both e1 and e2 are 'OFF', d1 pulls up the NOR OUTPUT to logic '1'. The second half of the XOR operates similarly, and a logic '0' is output if either e3 is 'ON', or if both e4 and e5 are 'ON'. Table 2.5 summarises the operation of this nMOS XOR gate.

Table 2.5 Operation of the nMOS XOR Gate of figure 2.14

A	B	e1	e2	e3	e4	e5	OUTPUT
0	0	off	off	on	off	off	0
0	1	off	on	off	off	on	1
1	0	on	off	off	on	off	1
1	1	on	on	off	on	on	0

The speed of operation of any nMOS logic gate is defined by the gate load capacitance, C_t, and the aspect ratio (W_d/L_d) of the pull-up transistor. The aspect ratio of the pull-down device is set by the need to obtain a correct logic '0' level. Thus a first estimate of aspect ratios would be obtained by calculating the gate load capacitance, C_t, and applying equation (2.13) followed by equation (2.12).

Introduction to MOS Devices and Logic

This information together with the value of the lateral under-diffusion for the process allows us to find the drawn widths and lengths of all the transistors used.

The value of 4 in equation (2.12) is valid for inverters or NOR gates. In a NAND gate with m imputs, a series string of m transistors replaces the single pull-down transistor of the inverter. To preserve the logic '0' level we must ensure that the combined effective current-carrying capability of all these devices when switched 'ON' is equal to that of the single pull-down device of the inverter. Thus the aspect ratio of *each* of the pull-down devices in an m-input NAND gate must be m-times greater than that of the equivalent inverter pull-down device. In other words, for an m-input NAND gate

$$\frac{W_e}{L_e} = 4 * m * \frac{W_d}{L_d} \tag{2.15}$$

For this reason, NOR logic is preferred to NAND logic in nMOS.

2.5 Synchronous Logic: An Introduction

A section of circuitry made up of logic gates and inverters is termed *Combinational Logic*. Signals propagate through sections of combinational logic as fast as the individual logic gates will allow. As more and more complex circuits are constructed, the use of purely combinational logic becomes untenable and it bcomes essential to discipline the movement of signals through the chip. If this is not done, circuits will be designed whose outputs depend on the *order* in which signals propagate, rather than purely depending on the inputs. This is clearly undesirable.

Furthermore, many circuit functions are either difficult or impossible to perform in purely combinational logic and it is necessary to introduce a *clock* signal to control data flow in the chip. The simplest form of clocking scheme is a 'two-phase non-overlapping clock' such as that shown in figure 2.15. While we are restricting our aims in this book to combinational logic, clocked logic is sufficiently important to warrant a brief discussion.

Figure 2.15 shows the fundamental principle of clocked logic. The registers transfer the logic values on their inputs to their outputs while the associated clock signal (1 or 2) is at a logic '1'. During clock(1), therefore, the inputs to the combinational logic blocks A and C are changing, and the outputs of registers b and d are held constant. During clock(2), the inputs to the combinational block B and D are changing, and the outputs of registers a and c are constant. Since clock(1) and clock(2) are non-overlapping, data move through this datapath structure in lockstep like boats through canal lock gates.

This disciplined data movement aids the designer of the combinational blocks. The time within which his logic block must operate is defined by the clock rate.

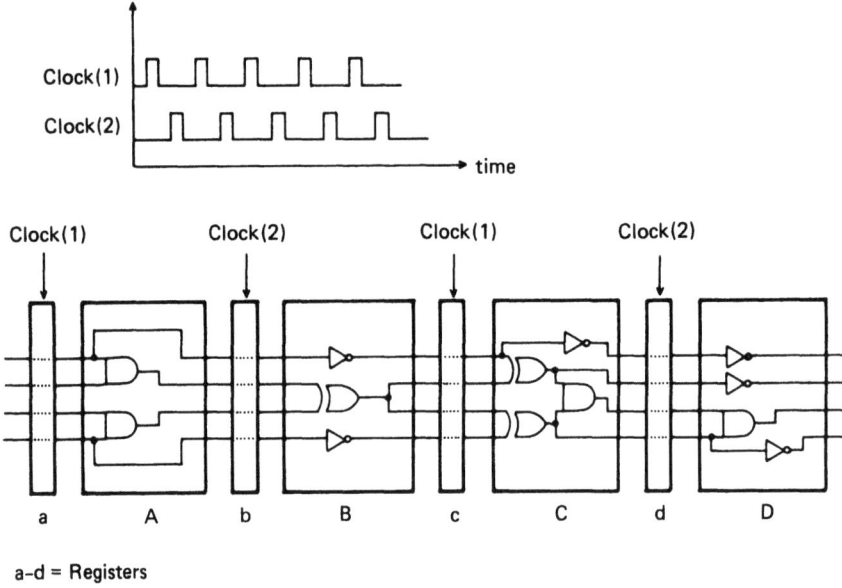

Figure 2.15 Two-phase clocked sequential logic (schematic)

As long as the overall function has been partitioned carefully into manageable blocks of combinational logic, this requirement should be easy to meet. This philosophy, which may be termed 'divide and conquer', permeates all aspects of IC design. It is much easier to design many small sections of circuitry which perform a complex function together than to attempt to design the whole circuit at once. With VLSI complexities, chips are designed by teams of engineers, and disciplined, well-partitioned chip architectures are essential.

2.6 Summary

This lengthy chapter has dealt with most of the important engineering aspects of MOS logic circuitry. The concepts will be related to a particular IC technology in section II. Many of the details included in chapter 2 can be filed in your mental archives. It is, however, important that you do not delude yourself by ignoring them, and believe that IC design can be carried out without reference to the underlying physics. Such an attitude may result in logically correct designs, but their speed and specifications will be unpredictable. That is not good engineering!

3 Fabrication of Silicon Integrated Circuits

Before an integrated circuit can be designed it is necessary to understand the *design rules* for the process on which it is to be fabricated. These design rules can be divided into three sections.

(1) The *geometric* rules which specify the minimum sizes of the shapes on every layer, the minimum separation between shapes, etc.
(2) The *electrical* parameters for the circuits, such as the transistor threshold voltage V_t, the transistor gain β, etc.
(3) Mandatory features required in mask-making and fabrication (such as markings on the lithographic masks that ensure accurate alignment of processing steps).

An integrated circuit designer should appreciate the reasons for the restrictions imposed on him by the design rules and therefore it is necessary for him to have an understanding of the processes involved in the fabrication of a silicon-based integrated circuit. To impart more than a basic understanding would be the work of several volumes and cannot be attempted here. There is no shortage of literature on this subject and the interested reader could refer to Glaser/Subak-Sharpe (see Bibliography at end of book) to pursue the subject one stage further.

We shall first discuss briefly the ideas behind the techniques used to make the raw material that is fundamental to integrated circuits – the silicon wafer. Most IC fabricators do not produce these themselves, preferring to buy them from outside companies who specialise in their manufacture. However, the designer should be aware of how they are made.

Once we have obtained our raw material, we must appreciate the methods used to form an integrated circuit from the silicon wafer.

An integrated circuit is created as a number of *layers*, each of which defines the areas on the chip to be affected by one process step. For example, all the metal interconnect for a certain design may be defined on layer number 7 and so for every rectangle drawn on this layer a rectangle of metal is produced on the chip. The rectangle is generated in terms of its coordinates from a computer database and this abstract data is converted into the actual rectangle of metal on the silicon wafer. This involves the use of *masks* and *photolithography* which are examined in sections 3.2 and 3.3, respectively.

Masks and photolithography allow us to define which areas of the chip will be subjected to any particular stage of the process. The major operations carried out on an MOS IC are oxidation, low pressure chemical vapour deposition (LPCVD), diffusion, ion implantation, and metallisation (interconnect). These operations will be examined in sections 3.4-3.8.

Finally we shall see how all these steps are combined to produce a working integrated circuit. The reasons for the design rules which are at the heart of every integrated circuit design will then become clear.

3.1 Wafer Production

The substrate or bulk material for an IC is usually a disc of silicon known as a *wafer* which is between 1 and 6 inches in diameter and about 0.5 mm thick.

To make the wafers it is necessary to produce a melt of pure silicon to which a controlled amount of the necessary impurities has been added. A seed crystal is brought into contact with the melt and is then withdrawn slowly, rotating all the time. As it is withdrawn the molten silicon cools and solidifies with the same crystal structure as that of the seed. The result is a near perfect single crystal of uniformly doped silicon (a boule) which is typically between 1 and 3 metres long and between 3 and 6 inches in diameter (though in 1985 sample wafers of up to 12 inches in diameter were available). As the seed crystal was inserted into the melt with a particular orientation, the crystal orientation of the boule (which will be the same as that of the seed) will be known, as will be the silicon 'type' (n-type or p-type). These are marked by grinding one or more 'flats' along its length. The boule is then sawn up into a large number of slices or wafers, which are cleaned and polished. These wafers are the 'raw material' for most integrated circuit fabrication lines. Few semiconductor manufacturers make them themselves, preferring to buy them from outside contractors. The cost of a wafer bought in bulk may be twenty pounds sterling but, as several hundred integrated circuits can be fabricated on a single wafer, this cost is a small proportion of the final cost of an IC.

3.2 Production of Integrated Circuit Masks

A mask consists of a transparent *substrate*, usually made of glass or quartz, upon which is placed a thin layer of an opaque, hard-wearing material such as chromium. This layer is etched selectively so that a rectangle from the design becomes *either* an opaque rectangle on a light background (light field mask) *or* a transparent rectangle on an opaque background (dark field mask).

One mask contains all the information to correctly process *one* layer of the chip, so a *set* of masks is required to make an integrated circuit. Each mask is covered by an *array* of chip designs (often all the same in a production environment) and thus represents a single layer for many chips.

It is not normal, except in cases where the finest possible resolution is required, to create masks directly from the design data. In production it is more common to create a *reticle* which is a mask for a single chip at either five or ten times its final size. The actual mask is made from the reticle by a further stage of photographic reduction.

To make a reticle it is necessary to start with a suitable base material, and coat it with a thin layer of chromium and then with a layer of photoresist, a material which is sensitive to light or an electron beam. This photoresist is then exposed selectively and developed, which leaves a stencil of the desired pattern. The complete reticle is then immersed in some chemical which will etch chromium but not the developed photoresist. The photoresist is then removed, leaving the desired pattern etched in the chromium. The selective exposure of the photoresist can be accomplished by either *optical* or *electron beam* techniques.

Reticles written using optical technology are produced by exposing to light a series of variable sized rectangular areas or 'flashes' on the reticle plate. The number of 'flashes' required to write the reticle plate depends on the complexity of the chip design. As a 'flash' takes a fixed time to expose, complicated designs take a long time to write and cost a lot of money.

The resolution of optical mask-making equipment is about 2.5 μm which is close to the physical limits set by the wavelength of light. However, the best optical equipment uses special techniques which can improve this value. If 2.5 μm resolution is available, and × 10 reticles are being written, then about 0.25 μm is the best resolution that can be expected on wafer (ignoring any loss of resolution when the reticle is reduced optically to × 1 size).

Optical mask-making technology and the photoresists used are well established and understood. The equipment has been installed and operators trained and thus there is a substantial investment in equipment and personnel. For these reasons alone, optical technology will be with us for some time.

If Electron beam (E-beam) techniques are to be used, an electron beam is scanned over the reticle in much the same way that an electron beam makes a picture in a television by scanning over phosphor. The only differences are that in the reticle writing process the resolution is far better, the beam is either ON or OFF (there are no shades of gray) and, of course, the beam writes only once on to photoresist, rather than many times on to phosphor.

E-Beam techniques are becoming more widely used in mask writing as the reticle write time is much shorter and the resolution is much higher than when optical techniques are used. This finer resolution (about 0.25 μm) can produce × 5 reticles which contain more than one chip design (making the creation of working masks quicker and cheaper) and also allows masks to be written directly without reticles. Because we can write masks directly, it becomes possible to place a number of *different* designs on the same mask, thus allowing the production of *multi-project* wafers – single wafers containing a number of different integrated circuit designs. These are extremely useful when working on prototypes, as a number of separate designs can be fabricated for the cost of only one mask set.

The patterns etched on reticles are not the same size as those to be made on the IC and so must be reduced to actual size. Also, very few copies of the chip design are contained on the reticle (usually only one) and we would hope to be able to fit many chips on to the mask and so on to the wafer. Thus we must expose the reticle repeatedly on to the mask plate until we have filled every possible position on the mask. This operation of size reduction and repeated exposure is called *step-and-repeat*. The mask created by the process can then be used to expose patterns on the silicon wafers.

3.3 Photolithography

The principle of 'lithography' was first developed in the late 18th century when patterns were transferred from a stone 'mask' to paper, thus allowing the cheap and easy reproduction of 'original' paintings. Needless to say, although the principle remains the same, lithographic fabrication of integrated circuits is rather different.

In the photolithographic stage of integrated circuit manufacture the wafer is coated with a photosensitive substance and light is shone *through the mask* on to the sensitised wafer surface. The coating is then developed, thus causing the pattern which was on the mask to be reproduced.

Before any photolithographic techniques can be applied, the wafer must be cleaned scrupulously and dried. It is then mounted on a vacuum chuck which spins the wafer at a speed of about 6000 rpm. A measured drop of *photoresist* is placed at the centre of the wafer which is then accelerated carefully to its full speed of rotation. This causes the photoresist to spread out uniformly and cover the entire surface of the wafer to a depth of about 1 micron. Photoresist is an organic substance, the properties of which change according to whether it has been exposed to ultra violet light (or Electron beams or X-rays, depending on type). If a *positive* photoresist is exposed to light it becomes more soluble in a developer and if a *negative* photoresist is exposed to light it becomes resistant to attack by the solvent used as a developer.

Negative photoresists tend to swell as they are developed and thus can only resolve patterns of about three times their thickness. Positive resists do not swell and therefore can improve on this performance. This advantage may, however, be offset by the fact that a given thickness of negative resist is more resistant to attack by the chemicals normally used in the fabrication process than is the same thickness of positive resist. There are therefore four possible combinations of light or dark field masks and positive or negative resists. These combinations are shown in figure 3.1 which also shows how the same result can be achieved in two different ways.

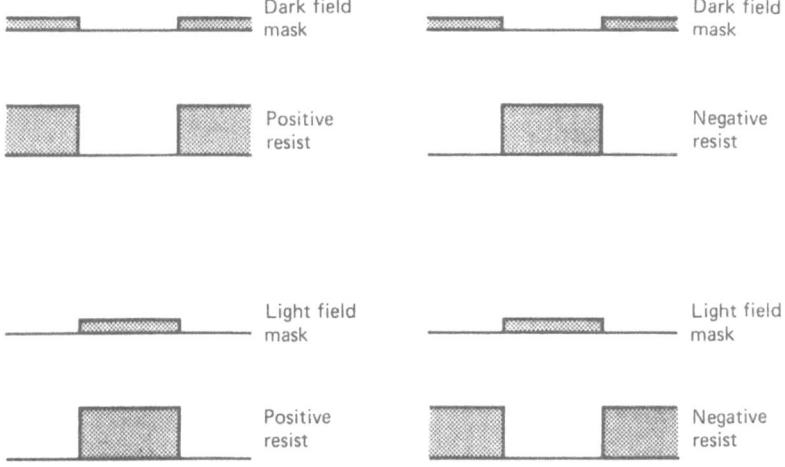

Figure 3.1 The four possible combinations of light and dark masks using positive and negative photoresists.

After the wafer has been covered in a thin layer of photoresist it is baked gently to drive out the solvent. Light is then shone through the mask on to the photoresist, using either a contact printer or an optical projection system.

In a contact printer the mask is brought into contact with the wafer and light is shone through the mask on to the photoresist. An optical projection system is rather like a photographic slide projector in that the slide (mask) is not in contact with the screen (wafer).

A contact printer gives better resolution than a one-to-one optical projection scheme, correct mask *registration* is easier to achieve (it is easier to align one mask to the pattern left on the wafer by the previous mask), and it is also cheaper, but inevitably after several prints the masks become damaged and covered with pieces of debris such as old photoresist. Though the debris can be removed by careful cleaning, the chip yield obtained when the mask is used drops and may eventually drop to such a level that the mask must be discarded.

As making masks is expensive, the projection system, in which nothing comes into contact with the mask, can be cheaper in the long term. However, this can be offset by the high capital depreciation costs incurred when using expensive optical projection equipment. An optical projection printer can use masks that are ten times larger than the final integrated circuit (that is, a reticle) and reduce the size optically. This means that any slight imperfections in the original mask will be reduced in size on the wafer by a factor of 10. The disadvantage of this technique is that exposing a wafer through a mask is no longer a 'one-step' process but instead a 'step-and-repeat' process, in which the same image is transferred once for every chip on the wafer. This is obviously time-consuming and therefore expensive but the increased yield obtained when making small geometry devices often implies that this process, known as *direct-step-on-wafer* (DSW), is the only viable technique.

Developed and hardened photoresist will protect the areas it covers from the chemicals used to etch silicon dioxide, silicon nitride and aluminium, and will also shield underlying areas from ion implants (see section 3.6). Photoresist will not, however, withstand high temperatures and thus cannot be used directly in conventional diffusion processes which take place at around 1100°C.

3.4 Oxidation

In any electrical system insulators are necessary, whether they be PVC in plastic-covered wiring, fibreglass in printed circuit boards or simply air between the plates of an air-spaced capacitor. In an integrated circuit the need for a good insulator is filled by silicon dioxide (SiO_2). SiO_2 is also of critical importance in masking high temperature process steps where photoresist would be destroyed.

Silicon dioxide can be *grown* on a wafer by exposing the silicon to dry oxygen or steam at high temperatures, or it can be *deposited* by pyrolysing (burning) silane (SiH_4) in oxygen. For further details on oxide deposition see section 3.7.

Growing SiO_2 is much slower than depositing it, but the resulting oxide is of a higher quality. However, growing oxide consumes some of the silicon wafer and this can sometimes be a disadvantage. Silicon can be oxidised either with or without water being present. Oxidation without water simply involves heating the wafers in the presence of dry oxygen. The oxide produced is of a high quality but the growth rate is slow. In wet oxidation the oxygen being fed to the reactor has water added. This may be by bubbling it through a constant-temperature, high-purity water bath or by adding steam at a high or a low pressure. In all cases the resulting oxide is of a slightly lower quality than that produced using dry oxygen, but the rate of oxide growth is much higher. An oxide layer is often created by starting and finishing the process in dry oxygen and using wet oxygen for the intermediate stage. This provides a good compromise between quality and growth rate.

It is because SiO_2 is a very good insulator and can mask very high temperature process steps that silicon is so attractive for the manufacture of integrated circuits.

3.5 Diffusion

Diffusion is the operation in which the number of donor or acceptor atoms in selected sections of the wafer are altered. In other words, during a diffusion operation selected parts of the wafer are made more n-type or more p-type.

During a diffusion process the silicon exposed to the dopant may change from n-type to p-type, or vice versa. The diffusion operation consists of *defining* the areas which are to be exposed to dopant and then *diffusing* the required amount of dopants into the selected area.

Silicon dioxide is used as the masking material because photoresist alone cannot withstand the high temperatures involved. Therefore to carry out the diffusion operation correctly we must ensure that the areas which are not to be diffused are covered in SiO_2 and that the areas which are to be exposed to the dopant are etched clean of oxide. Figure 3.2 shows a suitable sequence of operations.

There are two major techniques available to diffuse impurities (dopants) into silicon and the behaviour of each depends on the value of a factor known as the *diffusion coefficient*, D the units for which are m^2 s^{-1}. The value of D is different for every dopant and also varies dramatically with temperature. Graphs for D for commonly used dopants at a variety of temperatures are given in figure 3.3. We will now examine the two major diffusion techniques separately.

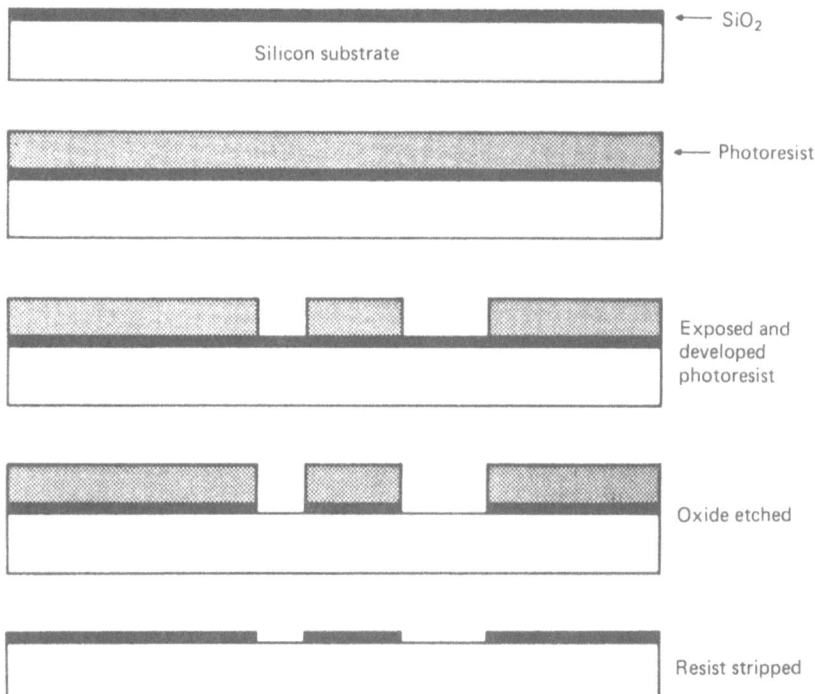

Figure 3.2 A sequence of operations that etches controlled mask shapes into the oxide layer

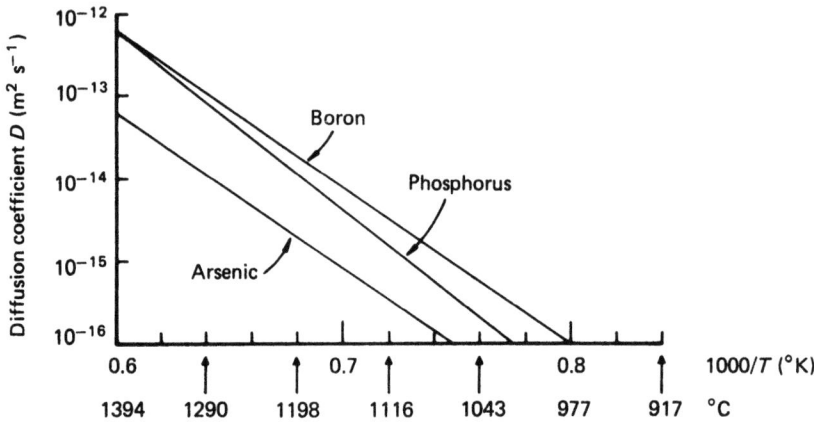

Figure 3.3 Diffusion coefficients for a variety of common dopants

3.5.1 Constant Source Diffusion

In this technique the dopant concentration at the surface of the wafer is kept *constant* at the value N_0 impurity atoms per cubic metre. The resultant diffusion profile $N(x, t)$, the impurity concentration at a depth x after a time t, is given by

$$N(x, t) = N_0 \, \text{erfc}[x/2\sqrt{(Dt)}] \text{ atoms/m}^3 \tag{3.1}$$

Values for erfc(z), the complementary error function, can be found from the graph given in figure 3.4. Figure 3.5 shows the diffusion profiles obtained by diffusing boron into silicon for a variety of times at a temperature of 1100°C. The constant source diffusion technique is often called the *complementary error function* diffusion because of the form of equation (3.1).

3.5.2 Limited Source Diffusion

Suppose that somehow a layer of Q atoms of impurities has been deposited over an area of A square metres on the surface of the wafer. Assume that it has somehow been arranged that the impurities so deposited cannot evaporate from the wafer and that no more impurities can be added. The source of the dopant is thus *limited*. The wafer is then heated to drive the dopant into the silicon.

The resultant diffusion profile will be

$$N(x, t) = \frac{Q}{A\sqrt{(\pi Dt)}} \exp\left(\frac{-x^2}{4Dt}\right) \text{ atoms/m}^3 \tag{3.2}$$

This profile is only exact if the pre-deposited layer was of zero thickness, but it is a good approximation to the true profile if $x < Dt/h$ and $t > h^2/D$ where h is

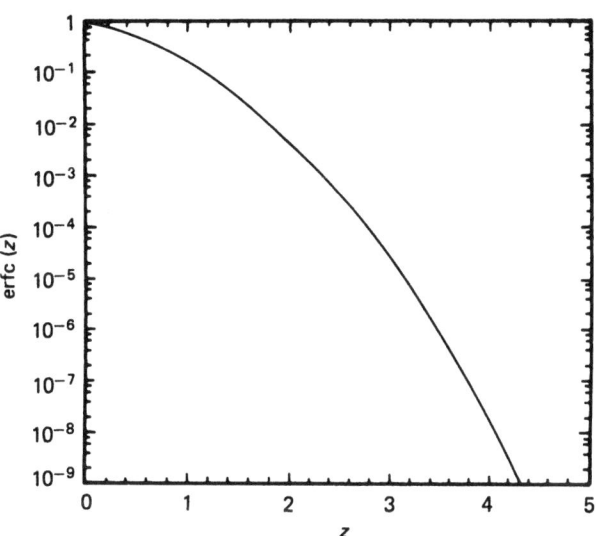

Figure 3.4 Functional form of erfc(z) (see text)

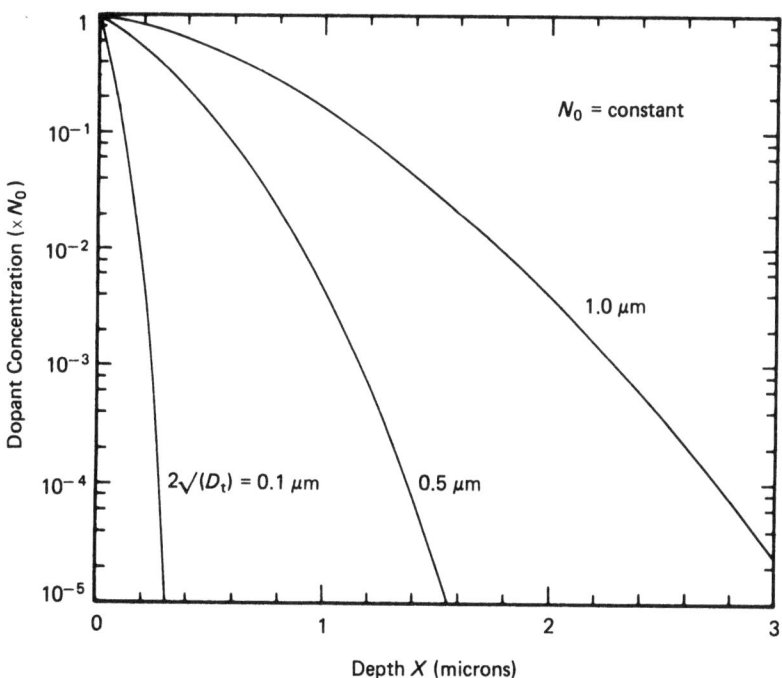

Figure 3.5 Diffusion profiles of boron into silicon for a variety of diffusion times (temperature = 1100°C)

the actual thickness of the pre-deposited layer. The limited source diffusion is also known as the *Gaussian diffusion* because of the form of equation (3.2).

The assumption that the impurities that have been pre-deposited on the surface will not evaporate from the wafer is valid in practice because while the diffusion takes place a layer of SiO_2 is grown on the wafer surface and this acts as a barrier to evaporation.

3.5.3 Practical Diffusion Techniques

A shallow, highly doped diffusion will usually be carried out using the constant source technique directly, but often a deeper, more uniformly doped diffusion is required and for this a combination of the constant source and limited source diffusion techniques is used.

First the wafer is exposed for a short time to a constant source diffusion with a very high surface concentration. This lays down a highly doped but very shallow diffused layer. This operation is called the *pre-deposition stage* and produces the thin surface layer of dopant that is required for a limited source diffusion. Obviously this layer is not of zero thickness but it will be extremely thin compared with the desired depth of the final diffusion layer. A limited source diffusion is then carried out, usually in an atmosphere of nitrogen and oxygen so that a thin layer of oxide will grow on the wafer surface to prevent the pre-deposited dopant from evaporating.

Before the effect of carrying out this limited source or *drive-in* diffusion can be predicted, it is necessary to find a value for Q/A. This can be cound by suitable manipulation of equation (3.1). The final value is

$$Q/A = 2N_0\sqrt{(D_1 t_1)}/\sqrt{\pi} \text{ atoms/m}^2$$

Substituting this value into equation (3.2) gives

$$N(x, t_2) = \frac{2N_0\sqrt{(D_1 t_1)}}{\pi * \sqrt{(D_2 t_2)}} \exp\left(\frac{-x^2}{4 * D_2 t_2}\right) \text{ atoms/m}^3 \qquad (3.3)$$

where D_2 is the value of the diffusion coefficient at the temperature, T_2, at which the drive-in diffusion is carried out. For $D_1 t_1 \leqslant D_2 t_2$, equation (3.3) adequately describes measured impurity profiles.

3.5.4 Typical Diffusion Apparatus

We know that the diffusion constant, D, is critically dependent on temperature, and that the higher the temperature the faster a given diffusion process will take place. In order to maximise the throughput of an IC fabrication plant it is obviously desirable to carry out the diffusion operation as fast as possible, consistent with the over-riding requirement of maintaining control and not degrading yield.

A diffusion temperature of less than about 700°C would result in diffusion times that would be unacceptably long for all but the most shallow junctions in extremely small geometry devices. The actual temperature used depends on the level of doping and the junction depth required but for medium size geometry devices and conventional diffused junction depths temperatures of over 1000°C are used. For VLSI (small geometry) devices with shallow junctions, temperatures of around 850°C are more usual. Whatever temperature is finally decided upon the furnaces in which the diffusion operations are carried out must hold this temperature to within ±0.5°C over their complete working area − no mean feat!

Gaseous dopants are heavily diluted in an inert carrier gas, such as nitrogen, and then passed over the heated wafers. If the dopant is in liquid form an inert carrier gas, such as nitrogen, is bubbled through it and carries it to the wafers. To allow a better control of dopant surface concentration it is usual to allow only a small percentage of the gas going into the furnace to bubble through the liquid dopant. Figure 3.6 shows a schematic representation of typical apparatus suitable for the diffusion of a liquid dopant, such as $POCl_3$. Solid dopants are frequently used as they allow the diffusion process to be more precisely controlled. They are often made in the form of wafers which can be placed between each of the silicon wafers in the furnace.

If a drive-in diffusion is taking place, it is necessary to pass oxygen into the furnace. This causes a 'capping' layer of SiO_2 to form, which prevents the dopant laid down during the pre-deposition stage from leaving the wafer.

3.5.5 Lateral Under-Diffusion

It is tempting to think of the direction of travel of dopants as being only down into the wafer but, unfortunately, they also diffuse sideways. Thus if a certain area is defined by some form of masking material and a diffusion process is carried out, the actual area in which dopant ions are present will be larger than that defined, by some process-dependent factor known as the *lateral under-*

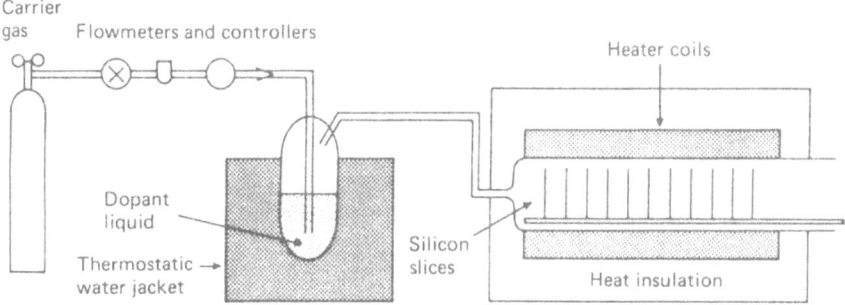

Figure 3.6 Diffusion apparatus for impregnating silicon wafers with $POCl_3$ (in liquid form)

diffusion. A typical value for this parameter for a 5 μm nMOS process is 1.25 μm. In an attempt to compensate for this it is usual to make the shapes on the diffusion mask slightly smaller by this same factor, which reduces the 'error'. This is, of course, only possible if the edge of the diffused area is defined by a shape on the diffusion mask. Unfortunately, this is not always the case because modern MOS ICs are fabricated on a *self-aligned gate* process in which the edges of transistor sources and drains are defined not only by the diffusion mask but also by the edge of the polysilicon gate. This has a major effect on the design procedure because this edge actually defines the length of a transistor. Thus the source and the drain of a transistor fabricated on such a process are a little larger than drawn, thus implying that the distance between them, the transistor length, is a little smaller than drawn.

Consideration of the points mentioned above shows that the actual length of an MOS transistor fabricated on a self-aligned gate process will be less than the drawn length by *twice* the lateral under-diffusion factor.

The self-aligned gate technique is used because it causes the transistor gate to be *automatically* aligned to the source and drain, despite mask registration errors. The fabrication steps used in such a process are given at the end of this chapter.

3.6 Ion Implantation

While the diffusion operations discussed in the previous sections are vitally important in the production of integrated circuits, there are serious constraints placed on the process designer if diffusion technology is used exclusively.

(1) The surface concentration after any diffusion operation must be higher than the doping background of the parent material.
(2) If multiple diffusions take place then the silicon is changed from n-type to p-type (or vice versa) more than once and the total number of impurities present increases to a point where carrier mobilities and lifetimes suffer. Because of this only two (or perhaps three) diffusions can be carried out on the same area of silicon.
(3) Only two types of impurity profile can result from a diffusion operation — either approximately Complementary Error Function or approximately Gaussian.

The use of *ion implantation* overcomes these problems and allows new doping profiles to be produced. Ion implantation is a technique that is used to introduce impurities into a silicon wafer by bombarding it with a high energy beam of ions. Use of this technique can result in a 'layer' (actually a Gaussian distribution) of dopant being introduced just *below* the surface of the wafer.

The advantages of ion implantation are:

(1) It can be used to overcome all the disadvantages of a conventional diffusion operation.

(2) The number of impurity atoms implanted can be controlled precisely.
(3) Ion implantation takes place at low temperatures.
(4) A wide range of materials can be used as masks; even photoresist alone can mask an area from an ion implant.
(5) Any number of impurities can be introduced in any order.
(6) Impurity peaks can be placed below the surface of the wafer.

An ion implanter consists of, in sequence, an ion source, a magnetic mass analyser, an acceleration, focusing and deflection system, and a target (wafer) holder. It is possible to interchange the mass analyser and the accelerator but this can cause practical difficulties. The entire interior of the system is a vacuum chamber. Figure 3.7 shows a schematic diagram of an ion implanter.

Considered simplistically, an ion source creates a plasma of the element to be implanted and uses an electric field to extract from this plasma the ions required.

After the ions are produced by the source they are fed into the mass analyser. There, while they are travelling at a relatively slow speed, they are subjected to a strong magnetic field. Ions of the correct element, with the correct charge, are deflected by an amount which aligns them with the exit from the analyser. Ions with an unexpected charge or an incorrect mass, are deflected by an amount that results in them hitting the walls of the analyser and being absorbed. Thus only the ions with the correct mass and charge leave the mass analyser and enter the ion accelerator, which increases their velocity to give them kinetic energies in the range 10–300 keV. The beam current, which is a few milliamps, is a measure of the number of ions being delivered per second.

It is difficult to obtain uniform current density in ion beams of large cross-section and therefore a focused spot is scanned over the surface of the wafer. The focusing of the beam is carried out electrostatically and the beam is then bent slightly before it is passed to the scanning system. This 'bending', which is achieved using either an electric or a magnetic field, ensures that any particles

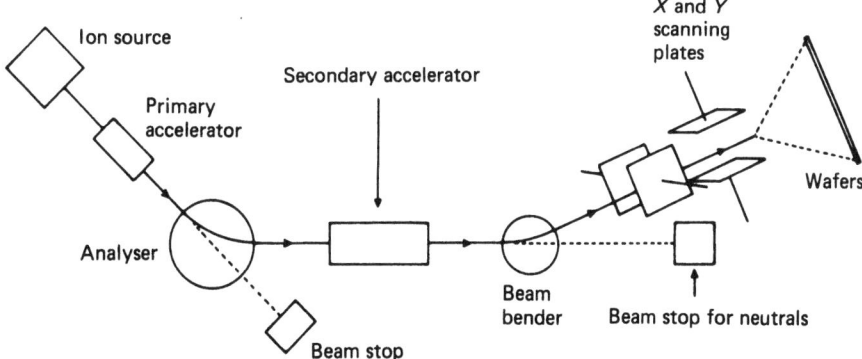

Figure 3.7 Ion implanter (schematic)

which have lost their electric charge (that is, are not ions) hit a 'beam-stop' and do not reach the wafer. Uncharged particles would not be affected by the scanning fields and would therefore be concentrated on one part of the wafer, rendering it useless. They also would not contribute to the electric charge being delivered to the wafer and thus their presence would go unrecorded.

After the 'beam-bender' the ion beam arrives in the scanning system which is either electrostatic or mechanical. In medium current machines electrostatic sweeps are most common but high current machines are now using mechanical or hybrid mechanical/electrostatic scans. In the past, mechanical motion of the target has been avoided because of the precision required and the risk of contamination by lubricants. However, as modern machines require larger loading capacity and faster throughput, target chambers now include complicated mechanics in the form of cassette loaders or carousels and therefore this objection to mechanical scanning has all but disappeared.

The final beam diameter is usually about 1 square centimetre in area and is either circular or rectangular, depending on the method of focusing. Its divergence will be about $0.1°$.

As the ions that are being implanted hit the surface of the silicon wafer, they lose energy by colliding with the silicon atoms. Such collisions are random and thus the final stopping point of the ions in the wafer is not precisely known. However, large numbers of ions are being implanted and so statistical methods can be used to find the implanted *distribution*. This turns out to be roughly Gaussian and can therefore be described by a mean, known as the *range*, R_p, and a standard deviation, known as the *straggle*, ΔR_p.

Ideally the wafers would be positioned such that the beam hits them at an angle of $90°$, but in practice they are tilted about $7°$ away from this position. The crystalline orientation of the wafer is such that a beam hitting it at $90°$ would be 'channelled' between the silicon atoms and travel further than expected in a rather unpredictable manner. If the wafers are tilted slightly the wafer looks like amorphous (non-crystalline) silicon to the beam, and the depth penetrated by the beam is better defined.

An ion implanter integrates beam current to the wafer and counts total charge. The dose can therefore be calculated accurately for amounts of 10^{20} ions/m^2 — equal to the heaviest diffused pre-deposition — down to less than 10^{15} ions/m^2. This is several orders of magnitude lower than what can be achieved by chemical diffusion processes, and ion implantation is also more accurate at any dose.

Dopants that have just been implanted are not incorporated into the silicon crystal lattice and are therefore not yet fully active. To activate the implant it is necessary to heat the wafer for a time to allow the dopant atoms to take up their correct position in the crystal lattice. Often this is achieved automatically in some later high temperature process step.

The most important advantage of the ion implanter over chemical diffusions is the accurate control over the amount and position of implanted dopant. In the chemical diffusion technique, the dopant profile is set by temperature, wafer surface cleanliness, and gas-flow rates. These factors are all much more difficult to control than voltage and current, the major variables in an ion implanter.

3.7 Low Pressure Chemical Vapour Deposition (LPCVD)

Layer deposition is a primary requirement of each cycle of the fabrication process. Apart from oxidation of the native silicon, all species must be transported to the surface where the film of required composition must be formed. The additional requirements are:

(1) Good adhesion.
(2) Uniform film thickness (usually < 3 per cent variation).
(3) Constant film composition.
(4) No aggregation.
(5) Good step coverage.

Chemical Vapour Deposition can be carried out in a cold-wall reactor at atmospheric pressure, in a hot-wall reactor at atmospheric pressure or in a hot-wall reactor at substantially less than atmospheric pressure. The first two of these systems are seldom used in the production environment today and we shall only consider the third — Low Pressure Chemical Vapour Deposition (LPCVD).

Layers of silicon nitride, polysilicon, silicon dioxide and some metals that meet all the above criteria can be deposited on large numbers of wafers using LPCVD. It has therefore played a crucial role in making the isoplanar silicon gate MOS processes viable for volume production.

LPCVD reactors are built into standard diffusion furnaces with slightly modified temperature controls. The system consists of a silica work tube designed for vacuum service, a gas control system and a vacuum system designed to accommodate the exhaust gases.

All the chemicals input to the reaction chamber are in gaseous form. If the system is operated at a low pressure ($\sim$100 mTorr) the mean free path length of the gas molecules increases to a value comparable to the chamber dimensions. Therefore, at a low pressure, the incidence of gas molecules is the same for any surface, no matter where it is within the chamber. Wafers may be stacked vertically for high throughput, but they still all see the same surface concentration.

Figure 3.8 shows a typical LPCVD reactor. The reactor walls are made of quartz or fused silica and are surrounded by heaters which heat the furnace tube to temperatures of up to 1000°C.

There are five main materials that can be deposited using LPCVD in a furnace of this type. These are silicon, polysilicon, silicon nitride, silicon dioxide and metals. Silicon deposition is not required in a basic MOS process so we shall not consider it further, but we shall briefly examine the role of LPCVD in the deposition of the other four materials.

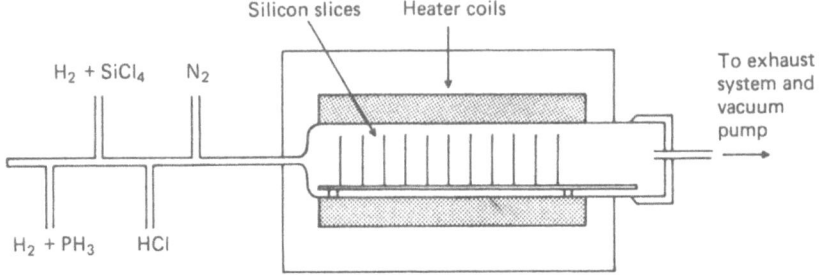

Figure 3.8 Low Pressure Chemical Vapour Deposition (LPCVD) reactor

(1) Polysilicon

Reaction is

$$SiH_4 \rightarrow Si + 2H_2$$

Polysilicon is one of the most successful LPCVD processes to be developed. Good quality and uniformity can be obtained across 150 wafers at a time, so the economics of production are excellent. Polysilicon is often used in its undoped state as a high value resistor, particularly for static RAMs. As an interconnect material, it is doped with phosphorus, thus lowering its resistance considerably.

(2) Silicon nitride

Reaction is

$$3SiCl_2H_2 + 4NH_3 \rightarrow Si_3N_4 + 6HCl + 6H_2$$

The performance of silicon nitride systems is comparable to that of polysilicon. They are most widely used to define active devices in LOCOS (Local Oxidation of Silicon) structures.

(3) Silicon dioxide

Reaction is

$$SiH_4 + O_2 \rightarrow SiO_2 + 2H_2$$

SiO_2 is used as an insulator between the polysilicon and the metal layer above. It must therefore be deposited at a relatively low temperature to avoid affecting the preceding process steps adversely.

(4) Metals

In some specialised applications aluminium, the normal metal used for interconnections on a chip, does not meet the requirements and harder metals such as tungsten must be used. LPCVD can be used to deposit these hard metals.

3.8 Metallisation

Using the processes outlined in the previous sections it should be possible to manufacture silicon gate MOS transistors. It should also be possible to interconnect the devices using either diffusion or polysilicon tracks. However, both of these materials have a high resistance (between 20 and 100 ohms per square) and a high parasitic capacitance to ground.

Both of these problems can be ameliorated if a metal, such as aluminium, is used for interconnections.

Before any metal layer can be deposited the entire wafer must be covered with an insulating layer of oxide to prevent any undesirable short-circuits. Holes are cut in this oxide layer to allow the required connections to be made to the underlying substrate and then the whole wafer is covered with a thin layer of the metal. This layer is then etched selectively to leave the desired interconnect pattern.

This is a very simplified view of the metallisation process and the interested reader should refer to Glaser and Subak-Sharpe (see Bibliography at end of book) for further details. However, there is at least one problem which should be mentioned here.

The holes that are cut in the insulating layer of oxide have sharp and steep edges, and when they are covered with metal there is a tendency for the metal to break or crack, causing poor connections and unreliable or inoperative devices. This is the problem of *step coverage*. The 'steps' on the chip which can break the metal are not only caused by the edges of contacts but also by the edges of active areas (diffusion) and polysilicon.

If the design rules are obeyed step coverage is a problem that can be left to the production engineers, but if the designer wishes to 'bend' a few rules step coverage is something which will have to be considered. As always, time spent consulting the fabrication people would not be wasted!

As chip sizes are decreased the current density carried by the interconnect tends to increase and we are now reaching the point where the most common metal used, aluminium, is not up to the job. Therefore there is a lot of interest in hard metals such as tungsten which can carry higher current densities, so allowing higher currents to be carried by narrower tracks. A widely used and simple process for the deposition of aluminium is *evaporation* but unfortunately the hard metals cannot be deposited using this technique and either *sputtering* or *LPCVD* must be employed.

The most important technical feature now required of new processes is the availability of two or more layers of metal. In this case the insulator between the two metal layers will not be SiO_2 (because the high temperatures used in the deposition of this material would destroy the first metal layer) but the problems of step coverage will be the same, or worse.

3.9 Example of nMOS Process

This book is mainly concerned with MOS design and thus it is appropriate that the process example examined should be MOS. The example that follows is that of an enhancement only, single level polysilicon gate, single level metal n-channel MOS process in which buried contacts (those directly from polysilicon to diffusion) are not available. This is an extremely basic process which would be little used in this form. However, it serves to illustrate how the major processing steps form a practical integrated circuit.

The process steps are as follows:

(1) **Initial clean.** Removal of any organic material and surface oxide in acid and hydrogen peroxide.
(2) **Initial oxide.** Production of a thermal oxide layer 0.05 microns thick. This provides thermal matching between the silicon wafer and the silicon nitride that will be laid down next. The thermal coefficient of silicon dioxide is between that of silicon nitride and silicon.
(3) **Silicon nitride.** Deposition of a thin film (0.05 microns thick) of silicon nitride by low pressure chemical vapour deposition (LPCVD).
(4) **Masking oxide.** Growth of an oxide layer by thermal oxidation of silicon on the silicon nitride surface.
(5) **Photolithography.** Definition of the regions required for active devices such as transistors. All areas not in this region will eventually be covered by a thick layer of oxide known as *field oxide*.
(6) **Boron implant.** Implantation of boron ions into the silicon surface. This is accomplished by ion implantation and the energy of the boron ions is such that they can penetrate the oxide-nitride-oxide sandwich but cannot penetrate the active regions, which are masked by the photoresist layer. Boron creates a net positive charge at the silicon surface between the areas that will become transistors, thereby providing electrical isolation between devices.
(7) **Oxide etch.** Removal of the thin oxide layer not covered by the photoresist coating.
(8) **Resist strip.** Removal of the photoresist in a strong oxidising agent such as oxygen plasma or fuming nitric acid.
(9) **Silicon nitride etch.** Etching of the silicon nitride in all areas not protected by the surface oxide. Photoresist cannot be used because it is attacked by the nitride etch chemicals.

(10) **Field oxide.** Oxidation of silicon on the wafer surface in wet oxygen yields a layer of oxide about 1 micron thick in all areas not protected by nitride. In nitride areas the resulting oxide thickness is only about 0.01 microns.
(11) **Oxide etch.** Removal of the thin surface layer of oxide produced on top of the nitride layer. This operation has little effect on the thick field oxide layer.
(12) **Nitride etch.** Removal of the nitride layer.
(13) **Oxide etch.** Removal of the thin oxide remaining in the regions that define the positions of active components.

The active areas, those in which the transistors will be made, have now been defined.

(14) **Gate oxide.** Growth of a very pure oxide film approximately 0.08 microns thick. This layer forms the basis of the MOS devices and its uniformity, purity and stability are critically important.
(15) **Boron implant.** Implantation of boron ions to define the operating voltage (threshold voltage) for enhancement-mode MOS transistors.
(16) **Resist strip.** Removal of any surface contamination produced by organic materials present in the ion implanter system.
(17) **Anneal.** Ion bombardment damages the silicon crystal structure. Annealing permits local atomic repair of the surface and removal of implantation damage.
(18) **Polysilicon deposition.** Use of an LPCVD process to create a polysilicon layer 0.35 microns thick. This polysilicon will be used as conductors and as the gate electrodes for transistors.

The transistor gate now exists.

(19) **Oxidation of polysilicon.** Production of a thin surface layer of oxide on the polysilicon surface by thermal oxidation.
(20) **Photolithography.** Definition of the polysilicon areas that are required for gate electrodes or interconnections between devices.
(21) **Oxide etch.** Removal of surface oxide in all regions not protected by photoresist.
(22) **Resist strip.** Removal of photoresist from polysilicon areas.
(23) **Polysilicon etch.** Removal of polysilicon in areas where it is exposed — areas not protected by the surface oxide layer. As with step (9), photoresist cannot be used as an etch barrier because it is attacked by the polysilicon etch.
(24) **Oxide etch.** Removal of oxide from the top surface of the polysilicon and from regions between the polysilicon and the field oxide.
(25) **Phosphorus deposition.** Diffusion of phosphorus into the polysilicon layer to increase its conductivity. The phosphorus also diffuses into exposed areas of the silicon substrate, thereby defining the other two electrodes of the MOS transistor — the source and drain regions.

Transistors have been formed. Note that the edges where the transistor source and drain meet the gate have been defined by the edge of the polysilicon, which was present before the diffusion operation took place. This is a self-aligned gate process (see section 3.5.5).

(26) **Phosphorus de-glaze.** Removal of the phosphosilicate glass grown during the last step. This material has a low melting point and is a relatively poor insulator.

(27) **Polysilicon oxide.** Growth of a layer of dense oxide 0.2 microns thick to provide a good base for the next step.

(28) **Pyrolytic oxide deposition.** Production of a uniform, well-defined phosphosilicate glass layer about 0.75 microns thick using phosphine gas.

(29) **Reflow.** Softening of the glass layer to smooth the sharp edges for improved step coverage.

(30) **Photolithography.** Definition of areas where contact will be made to polysilicon or diffused regions.

(31) **Reflow etch.** Removal of reflow oxide from areas not protected by photoresist to expose contact windows.

(32) **Resist strip.**

(33) **Second reflow.** Rounding of the edges of the contact holes.

Contact holes have now been made and we are ready to deposit the metal interconnect.

(34) **Aluminium deposition.** Evaporation of a layer of pure aluminium 1.5 microns thick over the whole wafer surface.

(35) **Photolithography.** Definition of aluminium interconnect pattern.

(36) **Aluminium etch.** Removal of aluminium from all areas not protected by photoresist.

(37) **Resist strip.**

(38) **Sinter.** Annealing at low temperatures to provide good contact between silicon and aluminium layers.

Interconnect layer now complete.

(39) **Pyrolytic oxide.** Deposition of a layer of oxide 0.75 microns thick over the whole wafer to provide mechanical/chemical protection of the finished circuit.

(40) **Photolithography.** Definition of oxide windows over bonding pads.

(41) **Pyro etch.** Removal of pyrolytic oxide over bonding pads to expose metal pads.

(42) **Resist strip.**

Wafer now ready to be wafer tested ('probed'), cut into individual chips ('diced'), packaged and fully tested before sale.

4 Design Rules

The designer of an integrated circuit must understand and adhere to a set of *Design Rules* that describe the constraints on his design freedom imposed by technological considerations. This chapter describes some typical design rules, and gives reasons for their existence.

4.1 Contents of Design Rules

Design rules can be divided into three parts – *geometric*, *electrical* and *mandatory features* – and we shall consider them separately. The information they contain is, however, often inter-related. For example, the minimum transistor size (geometric rule) is one of the factors which determines the maximum power supply voltage (electrical rule).

4.1.1 Geometric Design Rules

Geometric design rules specify minimum sizes of features and patterns and also the spacing between features. These rules are determined by the following factors:

(1) Mask registration – how well each mask is aligned to the wafer.
(2) Sideways diffusion (in the diffusion operation).
(3) The position of the layer being fabricated in the Z direction, that is, the 'height' of the layer.
(4) Control of photoresist exposure.
(5) Distortion of the silicon wafer (run-out).
(6) Available control on etching.
(7) Allowable electric field strength.

Let us consider each of these effects separately.

(1) Mask registration

As each layer of an integrated circuit is built up during the fabrication process we must ensure that it is correctly aligned to the layers that went before. However, if we align each layer to the previous layer we are obviously going to risk cumulative errors. For example, if the registration was correct to the nearest micron, the accumulated error from the first layer to the nth layer could be as much as $(n - 1)$ microns. We must align the first layer to the wafer itself and then align subsequent layers as far as possible to that first layer, thus avoiding a cumulative chain of errors. This is the technique used in practice, though there are some differences in the later stages when alignment to the previous layer is more appropriate. An example of a simple nMOS process is as follows.

Mask 1 (active area) is aligned to the 'flat' ground on the edge of the wafer.
Mask 2 (implant) is aligned to layer 1.
Mask 3 (buried contact) is aligned to layer 1.
Mask 4 (polysilicon) is aligned to layer 1.
Mask 5 (contact windows, metal/active area, metal/polysilicon) is aligned to layer 1.
Mask 6 (metal for interconnect) is aligned to layer 5.
Mask 7 (protective oxide) is aligned to layer 6.

The last two layers, 6 and 7, are not aligned to layer 7. The metal defined on layer 6 is aligned to the contact windows defined on layer 5 because only its position with respect to that layer is important in this nMOS process.

Layer 7 is a simple passivation mask which causes the whole circuit, with the exception of the bonding pads, to be covered with a glassy material to protect it from damage. As only the registration to the bonding pads is important, and these are part of layer 6, layer 7 is aligned to that layer.

(2) Sideways diffusion

The mask for the diffusion operation is produced by covering the wafer with a layer of SiO_2 which is then etched away in the areas which are to be doped. When the wafer is then heated in a furnace the dopant will not diffuse through the SiO_2 and the areas which are covered by this oxide layer should remain undoped. However, the dopant diffuses not only downwards through the etched holes in the layer of masking oxide, but sideways under the oxide, typically by about 80 per cent of the diffusion depth. Thus areas of diffusion are always a little larger than they are drawn on the mask. An attempt is made to compensate for this effect by adjusting the sizes of the shapes on the masks, but such corrections are never perfect and an allowance has to be made for this source of error.

(3) The effect of 'height' of layer

Some of the later layers, notably the contact windows and metal, are affected adversely by their position in the Z direction on the wafer. For example, contact windows are defined between metal and polysilicon or between metal and active area (diffusion). A contact window through the oxide layer above polysilicon will be at a different height from a contact window though the oxide layer to an underlying diffusion, and thus the image of the mask on the photoresist cannot be perfectly in focus in both cases. For this reason there is a limit on the achievable resolution for layers which are not all the same height.

(4) Control of photoresist exposure

This effect (and those following) are second order effects compared with the three mentioned above. If the photoresist is over-exposed the image will tend to increase in size, while if it is under-exposed it is possible that it will not harden (negative resist) or soften (positive resist) as required. This problem is now not very significant as control over the quality of photoresist and the exposure process have improved.

(5) Distortion of the silicon wafer (run-out)

During the fabrication of the wafer it is subject to a number of high temperature processes that cause the silicon to expand and contract. This results in an overall distortion of the wafer which causes the partly processed chips to be slightly out of position with respect to their neighbours. This implies that while it might be possible to position succeeding layers correctly on any one chip, it will not be possible to position the succeeding layers correctly on every chip.

This problem can be alleviated if a 'step-and-repeat' mask exposure system is used (see section 3.2).

(6) Available control on etching

The rate of etching depends on a number of factors, such as temperature, and unless these are well-controlled circuits may be over or under etched. The rate of etching is assessed by an examination of the etch control lines, a series of lines of ever decreasing width that must be present on every design.

(7) Allowable electric field strength

The separation between two points which are at different electric potentials must never become so small that the electric field between these two points becomes excessive. For example, in a 5 micron nMOS process the maximum length of a transistor is 5 μm and the transistor lateral under-diffusion (see section 3.5.5) may be 1.25 μm. Thus the actual transistor length may be 3.5 μm. If the maximum power supply voltage is 5 volts the electric field across the transistor will be (5 volts/3.5 μm) which is a substantial 1.4 MegaVolts/metre. While this particular value may not be excessive, a large enough electric field will cause the breakdown of the depletion region which blocks current flow. Thus, if we wish to reduce the size of the transistor we have to consider reducing the power supply voltage for the circuit.

4.1.2 Electrical Design Rules

The electrical design rules specify a number of electrical parameters that apply to circuits fabricated on the appropriate process. The maximum allowable power supply voltage, maximum and minimum transistor gains (β), transistor threshold voltages (V_t) (and spreads), the resistance per square of each of the layers and the capacitances between layers and to the substrate are examples of electrical parameters.

This information is usually given in the form of limits, rather than as exact values. For example, the gain for an enhancement transistor (β_e) might be given as 'between $20 * 10^{-6}$ A/V^2 and $30 * 10^{-6}$ A/V^2. The circuit designer must use 'worst-case' calculations to ensure that a circuit will operate for any value of β_e within the limits given. Such 'worst-case' design techniques are of great interest to the IC designer as they help him to maximise his circuit yield. Unfortunately, they are outwith the scope of this book but the interested reader can use Cluley (see Bibliography at end of book) as a starting point for further study.

4.1.3 Mandatory Features

The design rule set will include a list of mandatory features which must be present in every design. These include the etch control lines, alignment marks, identification marks for the design, specified scribe channel features and test structures.

In our experience it is not difficult for mis-understandings to arise between the circuit designer and the chip fabricator. The designer should always check his understanding of these features and confirm with the fabricator that this section of the design rules, in particular, is up-to-date.

4.2 Process-Independent Geometric Design Rules

The values for the parameters given in the design rules used by an IC manufacturer depend on many factors which vary from company to company. For example, the resolution of the machines available, acceptable yield and type of photolithographic process used will vary. The design rules used by one company

Design Rules

are often very different from those used by another. This is a problem for the IC designer who wishes to ensure that his design can be manufactured by more than one fabricator. To deal with this difficulty Mead and Conway (see Bibliography at end of book) proposed a set of *Process-Independent Design Rules* which specified the geometric design rules not in terms of absolute measurements but in terms of a parameter λ, the fundamental resolution of the process. Thus a design could be carried out completely in terms of λ and a value need not be assigned to this parameter until fabrication is about to begin.

The parameter λ is the maximum error in the distance between two points on the same or different layers after all variations due to processing factors have been taken into account and a safety factor has been added. For example, if two points on an IC were designed to be 20 μm apart but after all the inaccuracies due to processing were taken into account they could only be guaranteed to be betweeen 17 μm and 23 μm apart, λ would be 3 μm. Clearly the minimum line width (size of a feature) on any layer must be greater than λ. If this rule is not obeyed the feature may disappear during processing! In practice, 2λ is sufficient on nearly all layers.

In the next few sections we discuss the derivation of the process-independent design rules that are widely used today.

Diffusion

The minimum feature size on the diffusion layer is 2λ, as we have said. However, the spacing between two diffusion areas is slightly more difficult to decide upon.

Two diffusion areas only λ apart could touch after process inaccuracies had been taken into account, so clearly λ is not sufficient. A spacing of 2λ would give a minimum separation of λ after process variations, which may seem to be adequate. However, physics dictates that the important spacing is that between the depletion regions which lie around these diffused areas, not the spacing between the diffused areas themselves. This critical spacing will be less than λ and therefore 2λ spacing between the drawn diffusion areas is still insufficient.

We must therefore maintain a minimum spacing of 3λ between two unrelated diffusion regions. The design rules for diffusion are shown in figure 4.1.

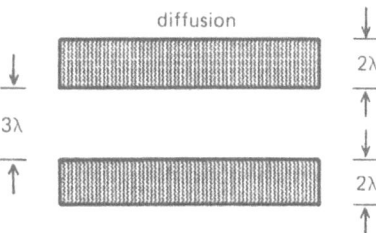

Figure 4.1 Technology-independent (λ) design rules for diffusion (active area)

Polysilicon

The minimum feature size for polysilicon is 2λ and, because there are no depletion regions associated with polysilicon lines, a spacing of 2λ between polysilicon shapes is sufficient.

If a polysilicon line and an *unrelated* diffusion line are parallel they should be separated by λ. If this rule is not obeyed, and as a result the diffusion area and and the polysilicon overlap, an unnecessary stray capacitance will result which will lower the circuit's operating speed.

If the diffusion and the polysilicon areas cross to form a transistor, we must ensure that the polysilicon crosses the diffusion area completely so that the transistor can be turned fully OFF. The polysilicon gate must therefore extend at least 2λ beyond the edge of the diffusion, as shown in figure 4.2.

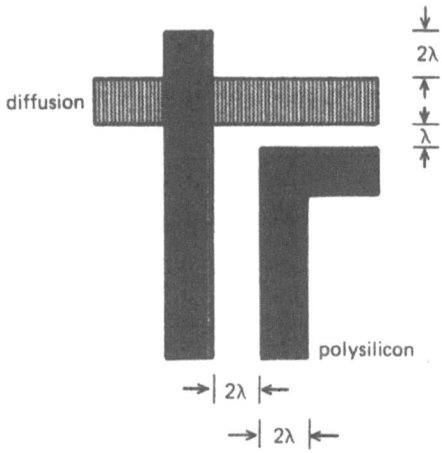

Figure 4.2   Technology-independent (λ) design rules for polysilicon

Ion implant

Transistors which are subjected to an ion implant are depletion devices while those which are not so exposed are enhancement devices, and so ion implant must not 'stray' on to an enhancement device. Furthermore, the complete area of a depletion device must receive its implant. The edge of the implant must therefore be at least 1.5λ from an enhancement device and an implant shape must overlap a depletion device by at least 1.5λ. In some processes these dimensions are increased to 2λ, and if a design is to be fabricated on several processes this higher value should be used. These rules are shown in figure 4.3.

Implant can be thought of as quite 'harmless' to all structures on the chip, with the obvious exception of enhancement transistors. For this reason implant

Design Rules 69

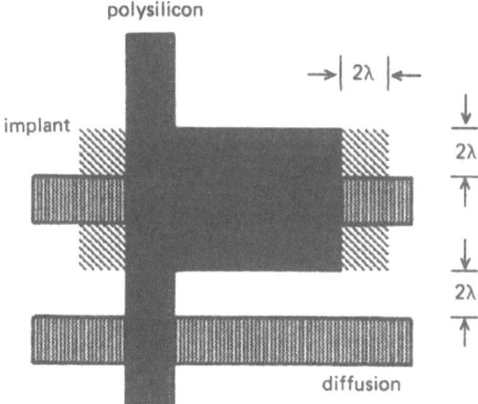

Figure 4.3 Technology-independent (λ) design rules for ion implant

is often placed freely, covering sections of the chip where its presence is pointless.

Buried contacts

The buried contact mask is used to remove the oxide selectively from beneath a transistor gate. Removal of this oxide causes the gate, source and drain of the final 'transistor' to be shorted together and a connection to be made from the polysilicon (gate) to the diffusion (source and drain).

Until recently, butting contacts were used to connect the polysilicon layer and the diffusion layer. However the use of butting contacts degraded chip yield and they are now unusual.

To make a buried contact the diffusion and polysilicon to be connected must overlap. This area of overlap must be at least $2\lambda \times 2\lambda$ in area. The buried contact must then overlap this area by at least λ all round. However, in the direction from the contact towards the diffusion this overlap is increased to 2λ. The reason for this extra overlap is demonstrated by the *incorrectly* designed buried contact shown in figure 4.4(a). After processing variations a buried contact intended to look like figure 4.4(a) may actually look like that shown in figure 4.4(b). The diffusion/polysilicon overlap area is *not* entirely covered by the buried contact and figure 4.4(c) shows the tiny transistor that is formed. This is not a direct contact and will probably cause the circuit to fail. To avoid failures of this type the overlap in the direction of the diffusion is increased to 2λ. If, in figure 4.4(a), the buried contact was to move to the left or right, a transistor would not be formed and a spacing of λ is sufficient in this direction.

Examples of correctly formed contacts are given in figure 4.5 Note that the diffusion and polysilicon areas in contact types (a) and (b) can 'move' with respect to one another during the chip fabrication without a reduction in their

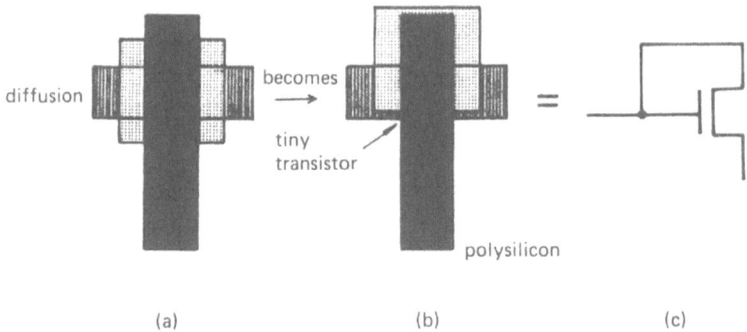

Figure 4.4 Incorrect buried contact

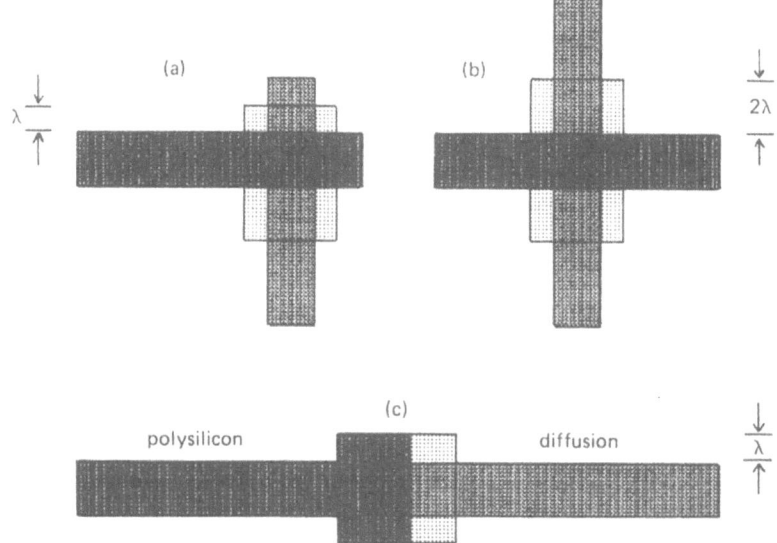

Figure 4.5 Correct buried contacts

area of overlap. Unless the area of the polysilicon is increased, this is not true of the contact shown in figure 4.5(c) and for this reason contacts of types (a) and (b) are preferred.

Buried contacts must be at least 2λ from any transistor, for obvious reasons!

In some processes, every buried contact must be included in the implant mask. Implanting buried contacts can decrease the contact resistance and so increase chip yield.

Design Rules

Contacts (metal to polysilicon or diffusion)

Contacts have a minimum feature size of $2\lambda \times 2\lambda$. Metal, polysilicon and diffusion must overlap a contact by at least λ. Contacts should be separated by 2λ and must be 2λ from any transistor.

If contact is being made between a large metal area and a large diffusion (or polysilicon) area, many small contacts should be used in preference to one large contact. These rules are shown in figure 4.6.

Metal

The metal layer lies on top of the diffusion and polysilicon layers and as a result crosses a surface that is far from smooth. More conservative design rules are therefore used for this layer. Minimum feature sizes and minimum spacings are usually given as 3λ. These rules are shown in figure 4.6.

Metal is commonly used for power wires on a chip and so is likely to be carrying a substantial current. If a metal power line is too thin, and the current density consequently too large, metal migration (movement of the metal in the direction of current flow), or even fusing due to evaporation of the metal, may occur. A 'rule of thumb' to prevent this allows one micron of metal conductor

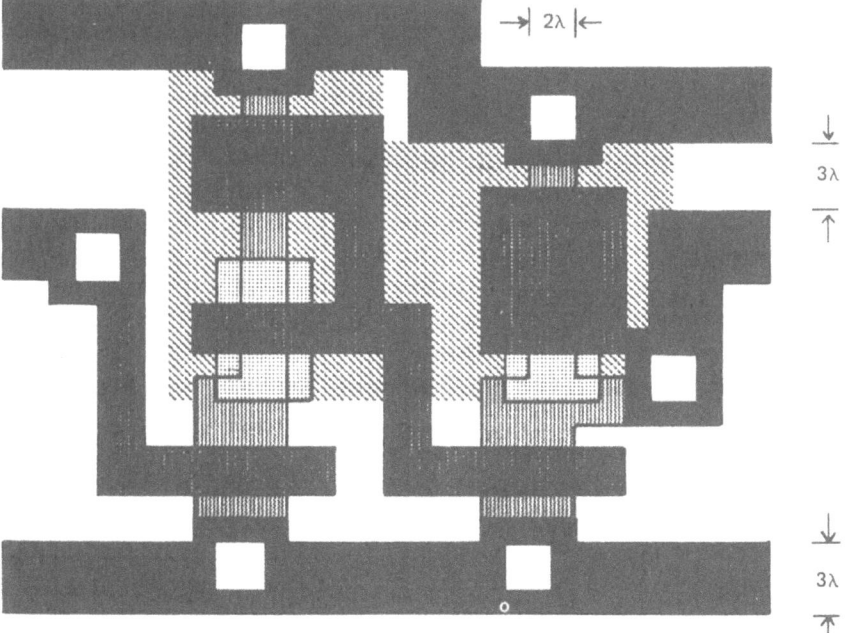

Figure 4.6 Layout (based on λ design rules) for two cascaded nMOS inverters

72   *Integrated Circuit Design*

width for every 10 mA of current flow. This rule will have to be altered if the actual value assigned to λ is reduced substantially below that used in the early 1980s.

4.2.1 Use of Buried Contacts to Define Transistor Gate Lengths

Figure 4.6 shows a layout for two cascaded nMOS inverters, each formed from an enhancement pull-down transistor and a depletion pull-up device. The latter has its gate connected to its source in the usual way for a depletion load. However, inspection of figure 4.6 will reveal that the gate-to-source connection has been formed differently in each inverter. In the left-hand inverter the connection has been formed in the 'obvious' way; a connection to the gate of the depletion transistor has been made through a polysilicon track and this track has been joined to the source of the transistor using a buried contact. In the right-hand inverter the gate and the contact have been 'merged', giving a more compact structure. Working down from the top of the right-hand depletion device in the diagram, the polysilicon first acts as the transistor gate and then moves straight into the buried contact area where it is connected to the transistor source. The disadvantage of this second technique is that when it is used the *length* of the transistor, a critical parameter, is defined by the final processed distance from the top edge of the polysilicon to the top edge of the buried contact and may therefore vary by as much as $\pm\lambda$ during processing. In the left-hand inverter the transistor length remains constant irrespective of process variations (within reason!).

The gate of the right-hand inverter in figure 4.6 is actually touching the diffusion that makes up part of the far-right metal/diffusion contact. This is apparently a violation of the rule that polysilicon and diffusion should be separated by at least λ (section 4.2). That rule is not being broken, however, because it applies only to *unrelated* polysilicon and diffusion. The diffusion and polysilicon areas that are touching in figure 4.6 are *related* because they are connected together.

4.3 Process-Independent Electrical Design Rules and Mandatory Features

Designs that can be fabricated on any process do not exist because electrical design rules and mandatory features are not constant from one manufacturer to the other.

The electrical design rules always vary from one process line to the other but, hopefully, the range of parameter values will not be too large. We can therefore expect that a *conservatively* designed chip can be fabricated on a range of processes.

The mandatory features often vary a great deal from one process line to the next. This problem can be avoided if the manufacturer supplies a pre-defined 'frame' into which the design is placed. This 'frame' is like a picture frame in that it surrounds the actual design, holds it all together and makes it usable. In IC fabrication, however, the 'frame' also contains all the required alignment marks, test structures, scribe channel details, etc. which are required to allow the chip to be successfully fabricated.

4.4 Advantages and Disadvantages of Process-Independent Design Rules

Integrated circuits designed using process-independent design rules have the following advantages when compared with those designed using absolute measurements:

(1) Several manufacturers may be able to fabricate the chip without modification to the design.
(2) As technology improves the absolute value assigned to λ at fabrication time can simply be reduced, thus producing smaller (and potentially faster) circuits.
(3) The λ rules are easier to memorise and need only be memorised once, thus making life easier for the designer.

Integrated circuits designed using the λ rules have the following major disadvantage:

(1) Slightly less circuitry can be packed into a given silicon area.
(2) An IC designer must have some idea of the value that will be assigned to λ when his chip is made. If an attempt is made to scale down the size of a design by assigning to λ a value much smaller than the designer expected, the current density in the metal interconnect will increase, perhaps to the point where failure occurs.

The λ design rules have been produced from the design rules (given in absolute measurements) of a number of manufacturers. They are therefore a compromise. No account can be taken of one particular manufacturer's use of, for example, equipment which allows reduced tolerances for one stage of the process. Thus if the parameter λ is used in an IC design, the resulting chip will take up slightly more area than it would if it had been designed for one particular process using an absolute measurement such as microns. In other words, silicon area is being sacrificed in order to ensure that the chip can be fabricated by a number of different manufacturers. Because of this increase in chip area, slightly fewer chips can be produced from a wafer of a single size and thus the cost per chip is increased, and the yield degraded.

The advantages of the λ rules are extremely important and they are used in much design work carried out today. However, most individual ICs sold today have been designed using absolute measurements. How does the designer know whether to use λ rules?

The main aim of good engineering is to make something which works well, sells well at a good price and makes a *profit*. Therefore the use (or otherwise) of λ rules must have something to do with profit.

The cost of an IC when it leaves the factory gates divides very roughly into two parts: the design costs and the production (fabrication) costs. The cost of designing the chip ends once the first working chip leaves the factory (excluding Mark 2, 3, ..., versions). Because this cost is spread over all the chips produced the design costs will become less important as the number of chips made increases. The cost of fabricating a *wafer* on a given process is constant, so the cost of fabricating a single chip is given by the wafer cost divided by the number of *working* chips on a wafer. A wafer can contain fewer large chips and there is also a smaller chance of them working (there is more to go wrong). This implies that the fabrication cost of a large chip is high.

An inefficiently designed chip will therefore have low design costs but high fabrication costs. The implications are as follows:

- If an IC is to be produced in small numbers, λ rules may be used to minimise the design costs (which will dominate). The final chip will be more expensive to fabricate but the small production run means that these costs are not so important.
- If an IC is to be mass-produced, the design costs are of less importance as they are spread over a large number of chips. The total chip cost will be dominated by fabrication costs which are minimised by expending effort (and money) on optimising the design. An optimised design does not use λ rules.

Therefore chips which are to be made in bulk are designed using absolute measurements, despite increased design costs, because the number of working chips on a wafer is thereby increased and profit is maximised.

4.5 Consequences of Breaking Design Rules

If the design rules are broken the result may be that the chip will not work or that its yield will be degraded. Following the design rules to the letter may be particularly important if λ rules are being used to ensure that the design can be fabricated by many different manufacturers. However, it is often possible to break the design rules without causing problems *if the designer knows what he is doing*. An examination of the design rules that can be broken, and when they can be broken, is outside the scope of this book and, in any case, it would be most unwise to discuss these details without defining exactly the process to which they apply. A good designer cultivates a friendship with the fabricators and finds from them the finer points of the process. Once he appreciates the nuances of the design rules he can figure out the consequences of 'bending' them.

5 Other Integrated Circuit Technologies

Thus far, this book has dealt exclusively with silicon MOS integrated circuits as this represents the majority of Application-Specific Integrated Circuits. However, the first semiconductor amplifier was a bipolar transistor and this device has not become significantly less important with the passing of the years. There is also considerable investment in a relatively recent technology known as 'Silicon-on-Insulator' (SOI). This is fundamentally an MOS process but the substrate is a wafer made of some insulating material. There are a number of materials which can be used for the substrate but the most promising to date is sapphire and, not unreasonably, this variant of SOI is known as Silicon-on-Sapphire (SOS). Silicon is not the only semiconductor known to science. There has always been germanium and, holding great promise for the future, there are the so-called III-V compounds such as gallium arsenide. These have some advantages over silicon.

In this chapter we shall look at some of these other technologies.

5.1 Silicon Bipolar Technologies

In small scale integrated circuits, bipolar circuit technologies have reigned supreme until the relatively recent arrival of robust CMOS circuits. This section is devoted to a discussion of the main bipolar logic families. It will help us in our attempt to understand them if we have at least a basic understanding of the bipolar transistor.

An NPN bipolar transistor is a sandwich of a thin layer of p-type silicon (the base) between two layers of n-type silicon (the emitter and collector). A PNP device is similar but has the doping types reversed. Though the details of the operation of a bipolar transistor are complex, the principle of operation is straightforward: a small current passed between the base and emitter controls the passage of a large current between collector and emitter. Note that in a bipolar device, current flows into the transistor control electrode (the base) whereas steady-state current should never slow into the gate of an MOS device.

In an MOS device there are three main modes of operation. The device can be 'OFF', in the linear region or saturated. We could also introduce a fourth mode of operation by interchanging the source and the drain, but as the device is symmetrical this is a little pointless. The bipolar transistor has, however, four distinct modes of operation and these are as follows:

(1) OFF. Both the emitter and collector junctions are reverse-biased. The current that flows is vanishingly small.
(2) Linear mode. In this mode the emitter junction is forward-biased and the collector is reverse-biased. To a first approximation the ratio of the base current to the collector current is approximately constant.
(3) Saturated mode. The emitter and collector junctions are forward-biased and there is a very low voltage (about 0.2 V) between the transistor emitter and collector. In this case a high density of minority carriers in the base region supports saturation. This charge must be removed before the transistor can be turned OFF. This takes time and limits the speed of operation of circuits containing transistors working in this mode.
(4) Inverted mode. In this mode the emitter junction is reverse-biased and the collector junction is forward-biased. The transistor works *backwards* with the functions of collector and emitter interchanged. Because bipolar transistors are not symmetrical they have a poor performance in inverted mode.

We can now discuss the main bipolar logic families.

5.1.1 Resistor Transistor Logic (RTL)

Resistor Transistor Logic was used in the construction of the first digital integrated circuits. An RTL NOR gate is shown in figure 5.1, from which it can be seen that its name is apt, as the circuit contains only transistors and resistors. When this logic family was first produced the 450 Ω base resistors were omitted and it was called Direct Coupled Transistor Logic (DCTL). However, when a gate output was connected to several DCTL gate inputs, the input transistors did not carry equal currents. This undesirable effect was reduced by the base resistors shown in the diagram. RTL has the most 'obvious' circuit of the bipolar logic families; transistors are connected in a common emitter configuration with resistors to limit the base and collector currents. If any gate input is connected to a logic '1', current flows into the transistor base. This causes a larger current to flow in the collector and the voltage on the gate output is pulled low. This logic family is slow because the transistor operates in the saturation region, and it has a low noise immunity and a low *fanout* (the number of loads that may be connected to a given output before performance is degraded). Despite being cheap and easy to fabricate, RTL is a thing of the past.

Other Integrated Circuit Technologies 77

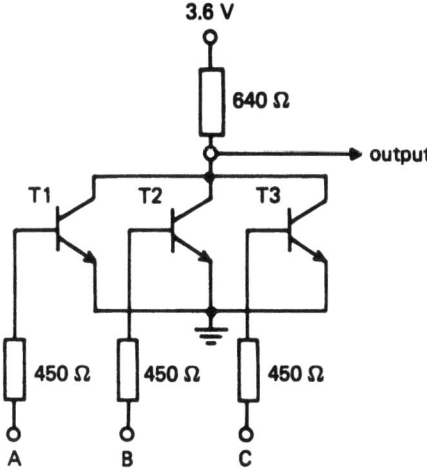

Figure 5.1 Three-input Resistor Transistor Logic (RTL) NOR gate

5.1.2 Diode Transistor Logic (DTL)

Shortly after the introduction of RTL circuits, Diode Transistor Logic devices were introduced. Figure 5.2 shows a DTL NAND gate. Improved DTL circuits, in which only one power supply was required, became available in later years. This logic family had a lower power consumption than RTL and also had the ability to deal with relatively large fanouts. However, it still suffered from the RTL problem of low noise immunity and had some extra problems of its own, in particular a high sensitivity to capacitive loads and temperature variations. Both RTL and DTL were limited in application and did not exceed the small

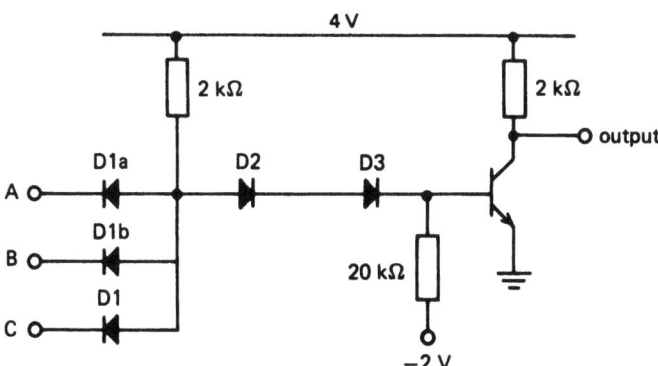

Figure 5.2 Three-input Diode Transistor Logic (DTL) NAND gate

scale integration level. They were completely superseded by Transistor Transistor Logic, a logic family which dominates SSI and MSI (Medium Scale Integration) even today.

5.1.3 Transistor Transistor Logic (TTL)

Figure 5.3 shows a TTL NAND gate. In TTL the logic '1' and '0' levels are often thought of as 5 V and 0 V respectively, but the actual specifications are as given in table 5.1.

Table 5.1 TTL Input and Output Voltage Levels

Gate Input	Gate Output
Less than 0.8 V	Greater than 2.4 V
Greater than 2.0 V	Less than 0.4 V

If any or all of the inputs to T1 are at logic '0' (less than 0.8 V) then, because T1 is turned hard 'ON' (and saturated) by the current flowing in its (4 kΩ) base resistor, the voltage on its collector will be low and transistors T2 and T3 will be 'OFF', allowing point Y to rise to 5 V.

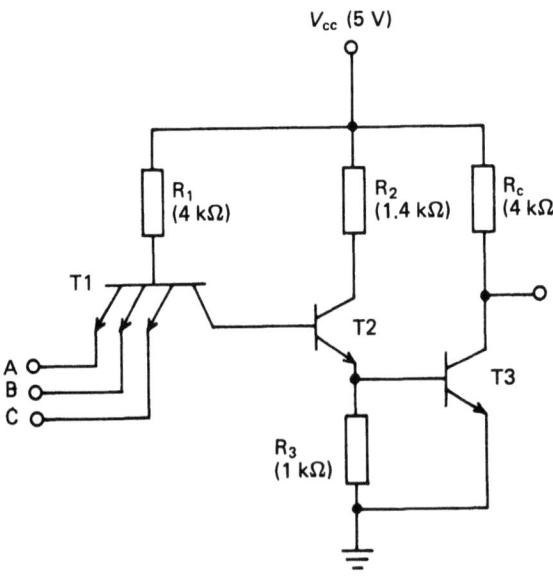

Figure 5.3 Three-input Transistor Transistor Logic (TTL) NAND gate

If all the inputs are at logic '1' (greater than 2 V) then transistor T1 will be inverted (will work *backwards*). This will allow current to flow *from* the gate input terminals to the collector of T1 and into the base of T2, causing T2 and T3 to saturate and pull the voltage on point Y to near zero.

More interestingly, when one of the inputs suddenly falls to logic low, the instantaneous voltage on the collector of T1 will not be low, though it will be falling, and T1 will still be in the linear region. It is therefore able to remove the charge quickly from the base-emitter regions of T2 and T3, allowing them to come out of saturation and turn 'OFF'. It is this transistor action which gives TTL the highest speed of operation of any saturated logic.

The basic gate in TTL is the NAND gate; NOR gates cannot be constructed directly.

Today so many variants on this basic TTL technology exist that 'standard' TTL is seldom used. The TTL family has expanded to include Low Power TTL, Schottky TTL, High Speed TTL and the latest introduction, Fast TTL. Usually the first two variants mentioned are combined to give Low Power Schottky TTL (the 'LS' series) while the other two are known as the 'H' and the 'F' series respectively. Because of the number of variants available it is hard to define the advantages of TTL, but all the types mentioned have a high noise immunity and can accept a high fanout. All of the TTL variants can operate at high clock rates, and the 'H' and 'F' series have particularly high operating speeds. Examples of TTL devices include counters, adders, multiplexers and flip-flops. Because of the dominance of TTL circuits they are produced in vast numbers and the resultant low price has led to their even more frequent use.

A major disadvantage of TTL is its high power consumption, particularly if standard TTL is used. CMOS still has a power advantage over every member of the TTL family at all but the highest clock rates, but the power consumption of the 'LS' series is comparable with nMOS. Both nMOS and CMOS have a great advantage over TTL in that they can be much more densely packed. Since the maximum size of a production chip is always limited, circuits of much greater complexity can be designed using these technologies than those produced using TTL. However, high speed TTL is much faster than either nMOS or CMOS.

5.1.4 Emitter Coupled Logic (ECL)

Emitter Coupled Logic is distinct from the bipolar logic families discussed thus far because it is an *unsaturated* form of logic. That is to say, the transistors in an ECL gate never enter the saturation region. This allows ECL logic gates to be extremely fast.

Figure 5.4(a) shows a basic *differential amplifier*. The power supplies for this circuit are unusual in that the positive side is at earth potential and the negative side is at -5.2 V. These are the conventional power supply voltages for ECL but we can mentally 'move' these voltages up and think of the positive rail as $+5.2$ V and the negative rail as 0 V.

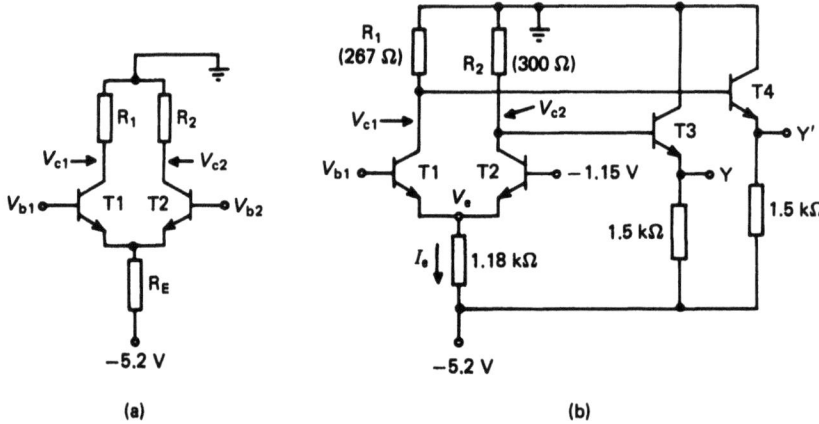

Figure 5.4 (a) Emitter Coupled Logic (ECL) gate differential stage. (b) Emitter Coupled Logic (ECL) inverter (for purposes of explanation only)

The sum of the currents in T1 and T2 is limited by resistor R_E and so if transistor T1 'asks' for more current, transistor T2 must receive less. If V_{b1} is significantly greater than V_{b2}, T1 will be turned hard 'ON' and will conduct all the available current, and no current flows through T2. The voltage V_{c1} falls (because of the increased voltage *across* R_1) and the voltage V_{c2} rises to zero volts (because there is no current in R_2).

If the V_{b1} input is connected to logic '½', a voltage level mid-way between that representing logic '1' and logic '0', and the V_{b2} input is connected to either logic '0' or logic '1', then the output V_{c1} rises if the V_{b1} input is a logic '0' and falls if the V_{b1} input is a logic '1'. This is how an ECL inverter works.

If we choose suitable values for the resistors, T1 and T2 are never saturated and the potential speed of operation of the circuit is increased.

In a practical logic gate the voltages representing the two logic levels on the circuit output must be the same as those representing the logic levels on the circuit input (so that we can cascade two such gates). Unfortunately, this is not so in the circuit shown in figure 5.4(a). The solution is to add a *level shifter* to the circuit output to give a final output that is always less than the differential amplifier output by a certain specified amount.

The circuit for an ECL inverter is shown in figure 5.4(b). Note that stand-alone inverters are not as common in ECL as they are in other logic families. This is because every ECL gate provides the true and the complemented version of its output and thus, within an ECL-based system, the logical inverse of every signal is already available. However, an inverter can be viewed as a one-input NOR gate and to change it into a three-input NOR, for example, it is only necessary to place a further two transistors in parallel with T1. The circuit analysis is unchanged.

The input stage of this gate is a differential amplifier with resistor values chosen to avoid transistor saturation. The outputs from the differential stage are taken to emitter followers which act as level shifters and also give the ECL gate a low output impedance. The circuit shown in figure 5.4(b) has values assigned to each of the resistors and while these values are not necessarily those used in any particular ECL gate, they will ensure correct gate operation.

Typical logic '0' and '1' voltage levels for an ECL gate are -1.55 V and -0.85 V respectively.

NAND gates cannot be constructed directly in ECL; only NOR gates are possible. However, because every signal in an ECL-based system is available together with its complement, use of **DeMorgan's rule** $[(\overline{A \cdot B}) = \overline{A} + \overline{B}]$ (where necessary) allows AND, NAND and OR gates to be made from the basic NOR gate.

The following points apply to ECL gates in general.

(1) None of the transistors will ever be saturated; this gate will switch extremely quickly and ECL gates can have propagation delays of less than 0.5 ns.
(2) The overall current drawn from the power supply is almost independent of the gate's output logic level. The power supply lines are therefore relatively free from switching noise.
(3) The voltage difference between a logic '1' and a logic '0' is only 800 mV which is small compared with other technologies. The noise margins (the difference between the output voltage levels achieved and the output voltage levels necessary for correct recognition of a logic level) are about 200 mV which, again, is small. Special techniques must therefore be used when laying out ECL components on a circuit board to minimise the effects of electrical noise. All voltages output from the circuit are produced with respect to the positive rail and therefore system noise is reduced if this rail is at earth potential.
(4) As in the circuit shown in figure 5.4(b), ECL circuits can be constructed to have both the gate output and its complement available. This eases the logic design.
(5) The low output impedance of the source follower stages allows the designer to interconnect ECL components using 50 Ω transmission line. Long, high-speed interconnections can therefore be made on a circuit board.
(6) Because of the low output impedance of these gates their fanout can be high (though it is limited by capacitive loading at high frequencies).
(7) The power consumption of ECL gates can be high but improved processing techniques are reducing this penalty.
(8) If ECL circuitry is to be interfaced to other logic families, special level shifters will be required.

ECL is the fastest silicon-based logic family available at present and its performance in terms of speed and power consumption is continually being improved.

However, it does not have the same packing density as either nMOS or CMOS and so is likely to remain an MIS, or at most LSI, technology.

5.1.5 Integrated Injection Logic (I^2L)

The circuits for the I^2L logic family can be derived from RTL but the actual layout of I^2L circuits on silicon is very different. The terminals of the different devices in the RTL gate are formed in one area of diffusion, which increases the circuit packing density until it rivals that of nMOS. Also the I^2L process is simple and can use as few as four masks.

The unique feature of I^2L is that the speed/power trade-off can be set by adjusting a single *off-chip* resistor. If we want more speed we can adjust the resistor to supply more current into the 'injector rail' which also causes a higher power supply current to be drawn. Conversely, of course, if we require less speed we can set the resistor so that less power will be used. We can even adjust the resistor value electronically so that we can introduce a 'standby' mode, allowing an I^2L system to be left switched on but drawing little power.

However, I^2L has some significant disadvantages. Because it is a *saturated* form of logic it is slow in operation. Also its high packing density can be equalled by CMOS which has the extra advantage that it consumes little power *all* the time. For these reasons I^2L has not fulfilled its initial promise.

5.2 Silicon-on-Sapphire (SOS)

One of the major reasons that the maximum operating speed of conventional MOS circuits fabricated on a silicon substrate is limited is that significant parasitic capacitances exist between each node of the circuit and ground. These capacitances are normally of no practical use but they must still be charged and discharged every time the circuit changes state. If these capacitances did not exist, circuit operation would be much faster. Circuits within a thin layer of silicon supported on an *insulating* substrate have been developed to solve this problem.

The substrate material used is *sapphire*, chosen because it is an insulator and also has a crystal structure similar to that of silicon. This latter feature of sapphire allows us to grow a high quality *expitaxial* layer of silicon on the surface of a sapphire wafer. MOS circuits are then fabricated within this silicon layer in much the same way as usual but with the additional step of *removing* the silicon from the areas where a semiconductor is not required. Thus the active devices are held within little 'islands' of silicon on an insulating substrate. A cross-section of an SOS IC is shown in figure 5.5. The presence of the insulator gives excellent isolation between individual transistors and also removes much of the parasitic capacitances normally present between each circuit node and ground.

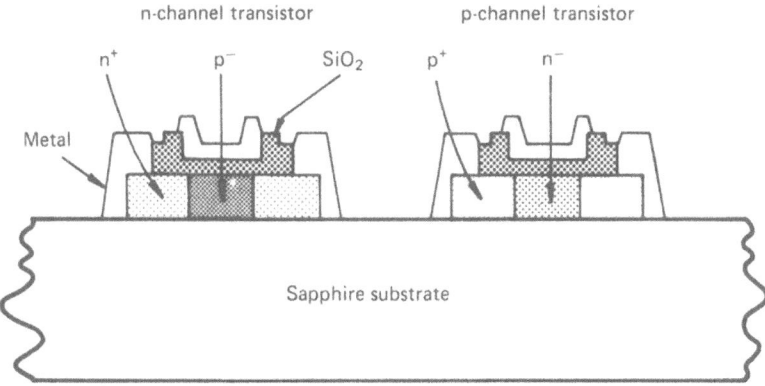

Figure 5.5 CMOS Silicon-On-Sapphire (SOS) process in cross-section

The only major capacitance remaining is the gate capacitance of each MOS device which cannot be eliminated.

An added advantage of this fabrication technique is that the resulting circuits can be exposed to substantial amounts of radiation and still work correctly – they are 'radiation hard'. This is obviously a great advantage in space and military applications.

SOS is significantly more expensive than conventional MOS circuits but is viable in applications where its extra speed and radiation hardness are important factors.

5.3 Gallium Arsenide (GaAs)

GaAs is a member of the group of compounds known as III-V semiconductors, so called because they are 'mixtures' of two elements, one from group III in the periodic table and one from group V. GaAs is not the only member of this group – others include indium phosphide (InP), gallium phosphide (GaP) and indium arsenide (InAs) – but GaAs is the III-V semiconductor most used today.

The useful III-V semiconductors have the great advantage that their electron mobility – the ease with which electrons can travel within them – is very high. Electron mobility in GaAs is about six times higher than that in silicon. Also undoped GaAs is a semi-insulator which, in the same way as in SOS, virtually eliminates circuit parasitic capacitances. The maximum operating frequency of GaAs transistors is therefore many times higher than in silicon. GaAs integrated circuits operating at clock rates of 2 GHz and above have been reported and there is little doubt that higher operating rates than this will be achieved in the future. Hole mobility in GaAs is less than that in silicon, so devices which use holes as charge carriers cannot be used to advantage in GaAs.

Apart from speed, GaAs circuits have other advantages. GaAs has a high band-gap energy which allows it to be employed in the manufacture of the LEDs and lasers used in interfaces with high-speed fibre optic links. Also because of its high band-gap energy, Schottky diodes fabricated in GaAs have a forward voltage higher than those fabricated using silicon-based technology. This is a big advantage in the fabrication of MESFETs (see below). GaAs is more radiation hard than any silicon-based technology (including SOS) and therefore GaAs-based ICs are finding applications in space and in military equipment.

The problems with GaAs lie in the nature of the material itself. Production of high-quality GaAs is difficult, GaAs has a low thermal conductivity, and it is brittle and likely to break during fabrication. However, the worst problem is the fact that the GaAs equivalent of silicon dioxide does not exist and no satisfactory substitute has yet been found. Therefore MOS transistors cannot be fabricated. We must use the MEtal Semiconductor Field Effect Transistor (MESFET), a FET with a metal gate which lies directly on the semiconductor. Such a metal/GaAs junction forms a Schottky diode which can be forward-biased, thus allowing current to flow into the transistor gate. Because of this current many design techniques which work well in MOS circuits (for example, dynamic logic) will not work if MESFETs are used.

A great deal of research is currently being carried out into GaAs and other III–V semiconductors. Most promising at present is the High Electron Mobility Transistor (HEMT), a MESFET device in which the semiconductor is made of extremely thin alternating layers of GaAs and gallium aluminium arsenide (GaAlAs). If integrated circuits composed of these devices could be made they would have clock rates of some tens of Giga-Hertz and would allow the construction of computers of immense speed and power. However, before that stage is reached the problems of fabrication technology will have to be overcome.

5.4 Comparison between Logic Families

The diagram given in figure 5.6 illustrates the *power–delay product* (measured in Joules) for each of the logic families discussed above. It is desirable to attempt to minimise the gate propagation delay while simultaneously minimising the circuit power consumption and so the 'best' logic families are placed near the bottom left of the diagram. This, however, takes no account of production cost or possible levels of integration and so is only one method of comparing these families. For example, TTL does not appear to be a very good technology from an examination of this figure but it is cheap, reliable and readily available and so it is still much in evidence. GaAs looks very good, being fast with a low power–delay product, but it is expensive, difficult to obtain and yield is such that only SSI/MSI circuits are available at present.

There is little doubt that CMOS is likely to dominate the mass production market in the coming years. It is low power, can be integrated to high levels and is robust and cheap. However, nMOS will still be used in a number of specialist, high density products such as memories. For very fast circuits GaAs (or other III-V compounds) is likely to come to the fore because of its advantages of high-speed operation, radiation hardness and (relatively) low power consumption. ECL will still be used in low-cost, high-speed applications where power consumption and radiation hardness are less critical.

Never before has the designer been faced with such a large number of choices of logic families from which to construct a system. Furthermore, the number of variants on each of these logic families is bound to increase, further complicating his choice. While this may not make the designer's life much easier, it should allow him to produce sytems with optimised speed, power consumption, radiation hardness, size and cost.

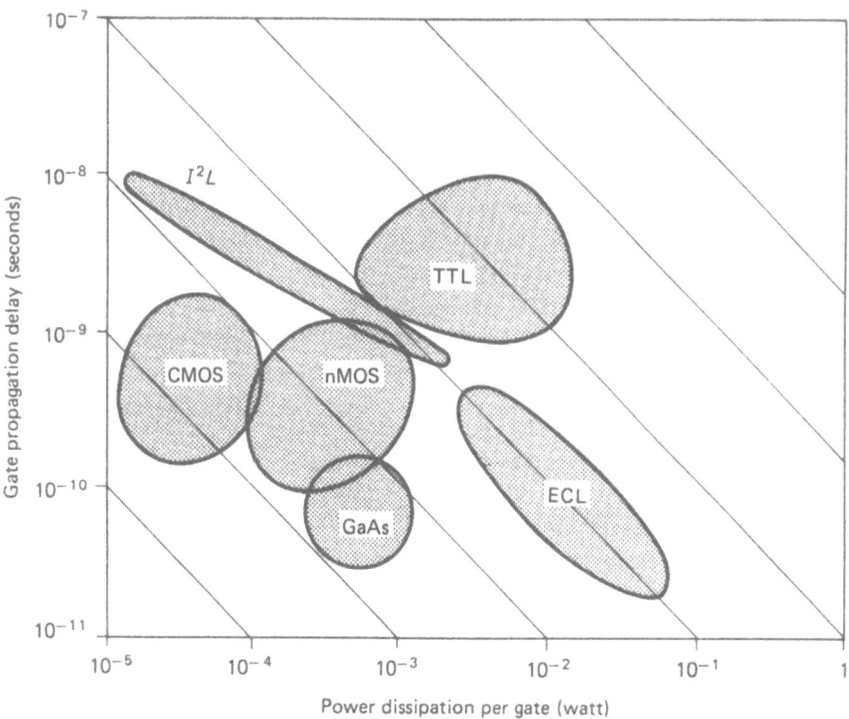

Figure 5.6 Power–delay continuums for VLSI circuit technologies

6 Integrated Circuits: from Concept to Silicon

In the first four chapters we discussed the reasons for using an Application-Specific Integrated Circuit in a system, and the principles behind ASIC technology and design. There are several distinct ways of turning the system requirement into an IC, each with its own advantages and problems. The main aim of this book is to take the reader through a simple design, consisting of only a few hundred transistors and wires, using only one of these 'routes', to a silicon IC. It would be short-sighted to use an exercise based on only one approach to design without at least glancing at the range of alternatives. This chapter aims to give an overview of the three distinct approaches to LSI/VLSI design, without delving into some of the more esoteric hybrid methods.

In addition, the use of Programmable Logic Array design, which is relevant to all three general approaches, is described.

6.1 Gate Array/Masterslice Design

The most conceptually simple approach, and that which will be used for the practical design example in section II, is that of the gate array or masterslice. It is also usually the cheapest route to silicon.

The starting point for a gate array design is a standard array of transistors on a prefabricated silicon substrate. The repeating unit in the transistor array is called a cell, and the designer must 'wire up' these cells to implement the desired function. This customising procedure is often referred to as *commitment*, as it commits a previously 'standard' array to a particular task. While few logic connections are present on a standard gate array before commitment, the power supply lines are almost always pre-defined, if not prefabricated.

Clearly the lower, normally unavailable layers of a gate array have to be designed in the first place and there are two schools of thought about the best topological style. There are some advantages to be gained by creating densely

Integrated Circuits: from Concept to Silicon

packed areas containing transistors, with intervening blank spaces for wiring. This is the 'streets-and-houses' school of thought. However, for some applications it is necessary to fill the silicon substrate with transistors, leaving no completely blank space for interconnections. This is often called the 'sea-of-gates' style of array. These alternatives are discussed in the following two sections.

6.1.1 Gate Arrays with Wiring Channels ('Streets and Houses')

Figure 6.1 shows a small section of a typical (fictional) gate array in which the transistor blocks are arranged in array cells separated by wiring channels. Interconnections can be made from polysilicon (short connections) or metal (longer wires). In general, gate arrays will have at least two levels of metal, thus enhancing the flexibility of interconnect, but in this example showing more than one layer of metallisation would only confuse the issue.

In using such an array to implement a logic function, the designer should first minimise the number of array cells (the basic array units) occupied by each circuit element (NAND gate, NOR gate, etc.). This involves small-scale design of the short-range transistor–transistor wiring. It is then necessary to optimise the placement of the circuit elements. Consideration should be given at this stage

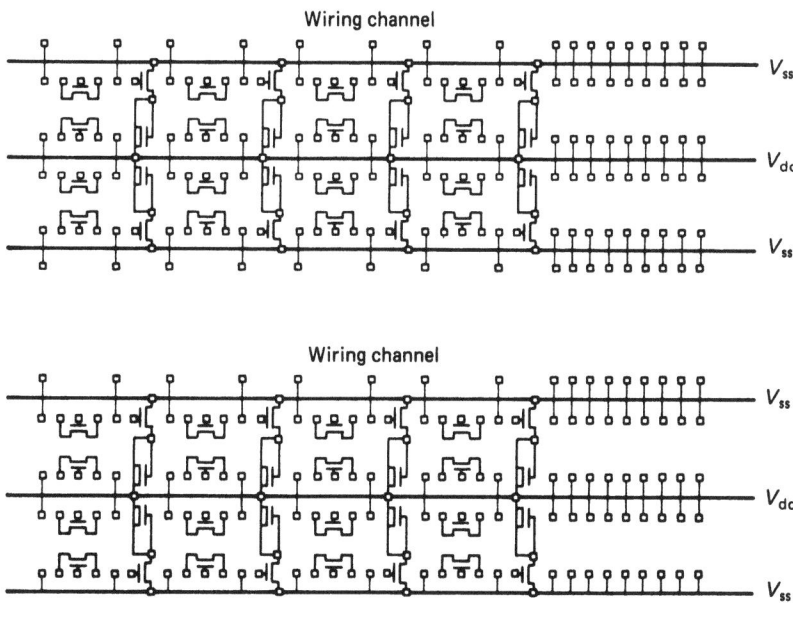

○────○ = underpass

Figure 6.1 'Streets-and-houses' gate array (schematic)

88   *Integrated Circuit Design*

to the connections *between* circuit elements. For instance, placing two elements that are connected together a long way apart will increase the interconnection problem unnecessarily.

Once the building block design and placement are complete the design is completed by the addition of the longer-range interconnections or *routing*.

The presence of the empty wiring channels should simplify routing, but dense circuitry can always lead to problems. The advantages and disadvantages of this style of gate array are as follows.

Advantages

(1) The structured form of the array (array cells separated by wiring channels) facilitates automation of the routing process, which is tedious and time-consuming if carried out manually.
(2) The density of the array cell areas minimises wastage due to unused local interconnections.

Disadvantages

(1) The pre-defined wiring channels may be too wide, and silicon may be wasted.
(2) The pre-defined wiring channels may be too narrow. If insufficient wiring space has been provided, rows of array cells will be left blank and silicon area may be wasted.

6.1.2 Sea-of-Gates Array

Figure 6.2 shows a section of the gate array that we will encounter in section II. There are no wiring channels, but there are more 'underpasses' to allow signal wires to pass under the power supplies. Superficially, the density of this 'sea-of-gates' array appears to be greater than that of the previous one with wiring channels but the topology of the array cell is actually substantially the same, and the cell repeat distances in the x and y directions are identical.

The design procedure is the same as for the previous style of gate array. However, the more open cell structure makes design of building blocks easier, while the lack of totally blank wiring channels makes routing more difficult.

In a 'sea-of-gates' array, rows of transistors have to be sacrificed when a horizontal or vertical wiring channel is required, when wires are simply run over the top of the transistors, which can no longer be used. This wastage of transistors can be minimised if circuit elements are designed to connect to one another by abutment. Elements designed in this way have their inputs and outputs so positioned that they connect together like 'Lego' bricks, with no explicit interconnect wires.

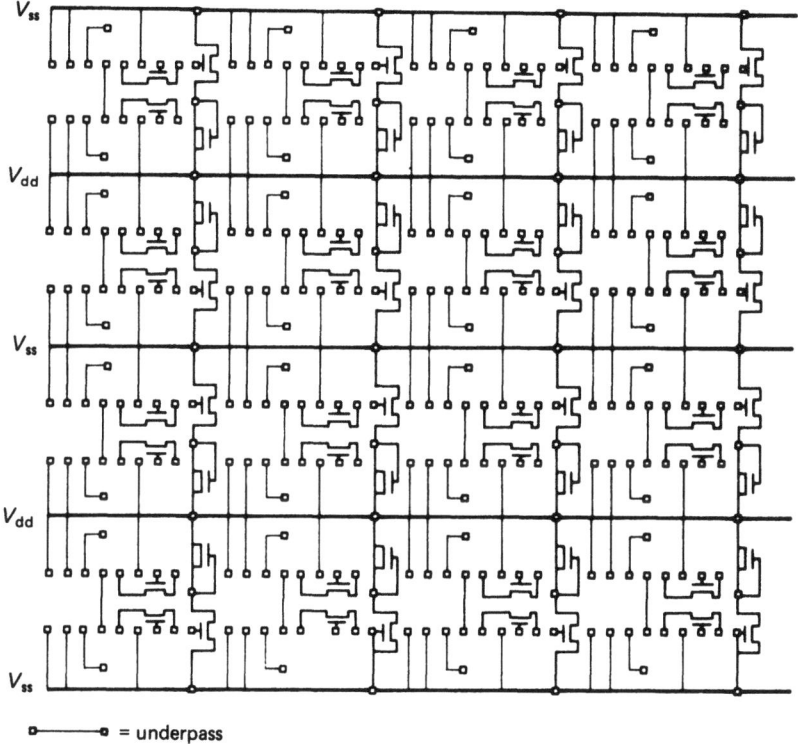

Figure 6.2 'Sea-of-gates' gate array (schematic)

It is not easy to see when connection by abutment is feasible, which means that automation is difficult with 'sea-of-gates' arrays. However, because of the evolution of better software techniques, the signs are that a resurgence in the use of 'sea-of-gates' arrays is occurring.

In all commercial gate arrays, the vendor provides a 'cell library' of predefined circuit building blocks, (such as NAND gates, etc.) which have already been fabricated and tested. This circumvents the need for the gate array designer to spend time designing compact circuit elements.

6.1.3 Gate Arrays: Advantages and Disadvantages

The major advantages of gate arrays are related to expediency and low cost. These are as follows.

Advantages

(1) There are fewer layers to be designed, so the time taken to design and produce a prototype is short.
(2) Fewer layers mean fewer lithographic masks. Once again, commitment is faster and cheaper.
(3) The existence of a fixed underlying structure makes design automation easier.
(4) Uncommitted arrays may be produced in large numbers, with all the savings that this implies.

The disadvantages are related to performance, and are summarised below.

Disadvantages

(1) The size of the uncommitted gate array chip is unlikely to be exactly right for the particular function desired. This means that the area will be inefficiently used, and the yield degraded.
(2) No fine tuning of the circuit's performance is possible, as transistor sizes cannot be changed.
(3) The fixed format of the underlying array structure means that some problems may be encountered which have solutions that cannot be realised in the form of a gate array design.
(4) Because there is a known underlying structure, and only a few layers of commitment, a design may be copied fairly easily by a knowledgeable plagiarist with a microscope.

Gate arrays are often used to condense circuit boards of random logic (usually TTL SSI/MSI parts) on to single chips. Because of this, many vendors aim gate arrays at TTL logic designers and offer 'cell libraries' containing pre-defined cells which correspond to standard TTL parts. In this application, the number of input/output pins on the array may be a greater limitation than the raw transistor count. To help solve this problem there are on the market some large arrays which have to be bonded into exotic packages, some of which have more than 100 input/output pins.

6.2 Standard Cell Design

The second route to silicon, which is generally more expensive than a gate array, is that of standard cell design. In this design technique the designer is presented with a blank piece of silicon, and a set of pre-defined circuit elements to place on it and wire together. There is no underlying array structure, although there may be restrictions on the freedom of placement of the pre-defined elements, which are called 'standard cells'.

The elements in a standard cell library may vary in size and complexity from a single gate to a complete microprocessor Arithmetic Logic Unit. If the standard cells are small functional blocks, they are versatile but tend to demand a large area of interconnect space. If they are large and complex, the internal density of each standard cell can be high but they are likely to be useful only in a few special applications. For instance, an ALU cell is useful to the designer who wants some microprocessor-like computing power on his chip but it is of little use to the designer who wants random logic. Thus the contents of the cell library will be a compromise between large, densely packed but perhaps over-specialised cells and small, versatile but less densely packed cells. The ideal cell library would have both types but, unfortunately, all commercial cell libraries are non-ideal!

An Integrated Circuit designed using a standard cell library may look like that shown in figure 6.3. The blocks of circuitry will be densely packed sections of full custom chip layout (see next section), while the wiring may have been carried out manually or automatically. There is an 'umbilical cord' of power supplies (V_{ss}, V_{dd}) and clock wires (ϕ_1, ϕ_2). This technique is often used to limit the variations on the chip 'floorplan' (the chip layout and routing strategy) as it simplifies automation of the cell placement and routing. Such constraints are less important if, as in figure 6.3, the cells are complex and large.

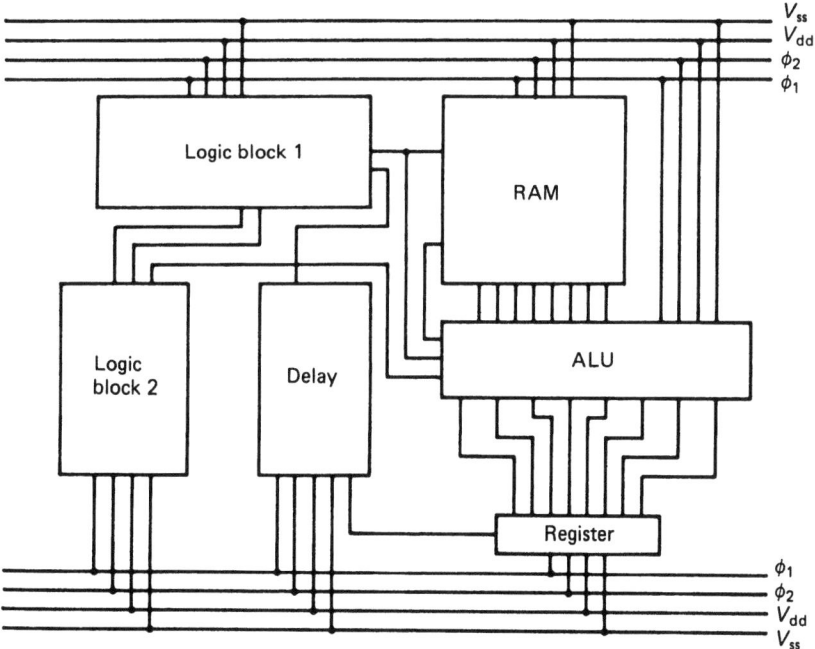

Figure 6.3 Section of a standard cell integrated circuit (schematic)

The elements called up from the standard cell libraries offered by some manufacturers are 'parameterised' by assigning values to one or more variables to change the form or the performance of the block. In this way, for example, while the cell library may contain only a one-bit shift register explicitly, the designer can call an eight-bit shift register, if required, by connecting together eight one-bit segments. It may be possible to achieve this with a single command to the CAD system of the form 'place shiftreg(8)'.

6.2.1 Standard Cell Design: Advantages and Disadvantages

Standard cell design is usually selected when it is necessary to achieve a higher level of performance and integration than is possible using a gate array. As with every design technique it has advantages and disadvantages.

Advantages

(1) ASICs designed using standard cells will have a good yield because the library elements will have been handcrafted to make optimum use of silicon area.
(2) The use of large sections of pre-designed circuitry means that performance can be closely controlled, that design proceeds quickly, and that the design is likely to be correct first time.
(3) As there is no underlying array structure, a standard cell IC may, in principle, be no bigger than the circuit area utilised.
(4) A standard cell IC is more individual than a gate array chip and is therefore more difficult to copy.

Disadvantages

(1) If a particular function is required that is not included in the standard cell library, and cannot be constructed from standard cells that *are* in the library, the IC cannot be designed. This restriction can be overcome if facilities are available to allow the inclusion of sections of custom design (see next section).
(2) A standard cell design requires a full mask set. This lengthens the time required to obtain a prototype and increases the development cost.
(3) A standard cell IC is only as good as the cells (and therefore the library) from which it is constructed. If the library is specified for clock frequencies of 20 MHz, and a 21 MHz clock rate is required, the chip cannot be designed (using this cell library).

Standard cell and gate array design facilities satisfy a large proportion of the world's need for ACICs. However, there are a significant number of applications where performance, design security and high yield are critical and the disadvantages outlined in the previous sections are significant. Examples of such applications are found frequently in military, medical and academic establishments. The last resort in such circumstances is to grasp the nettle of cost, complexity and design effort, and implement a full custom design.

6.3 Full Custom (Handcrafted) ASIC Design

When integrated circuit design and production was in its infancy, design was a manual process. The shapes of the transistors, wires, etc. were drawn on paper, transcribed on to and cut out of Mylar film, and subsequently reduced photographically to produce lithographic masks. A huge (carpet-sized) enlargement of the entire design artwork was then examined for design rule violations and other mistakes. This procedure was so tedious, error-prone and expensive that it is surprising that any ICs were actually produced.

Today's full custom design process is conceptually the same, but sophisticated Computer Aided Design (CAD) tools are used to reduce the difficulty and increase the accuracy. The nature of the CAD tools used is described in chapter 7. To specify the positions of the silicon, polysilicon, metal, implants, etc. that make up an IC may require the production of up to 13 masks. The IC designer, armed with a set of design rules such as those described in chapter 4, has to form the patterns on each of these layers. In doing so, he will look for simplifications such as regularity and repeated structure, but will still be faced with the task of designing several million shapes correctly, accurately and without violating design rules. The CAD tools will help, but the level of complexity involved means that few true VLSI chips have a single designer, and that the design process is a team effort.

Figure 6.4 shows a small section of a VLSI design. The CAD tools used to produce this piece of design are based on a microcomputer workstation which uses interactive colour graphics but in this figure different forms of crosshatching have been substituted for colour. This is not totally satisfactory, but it does keep the cost of this book down! The details are not of great importance, but it is clear that care is required in the placement of the overlapping shapes which make up the circuitry. In a simple CAD system for full custom design, it is the designer's responsibility to ensure that he does not break any design rules when placing shapes, but in more sophisticated systems, the CAD tools will warn of design rule violations. This will be taken up in the next chapter.

The advantages of full custom design relate to the ability to optimise the IC and are as follows.

Advantages

(1) The handcrafting that goes into the design means that use of silicon area may be optimised. Yield can therefore be maximised.

94 *Integrated Circuit Design*

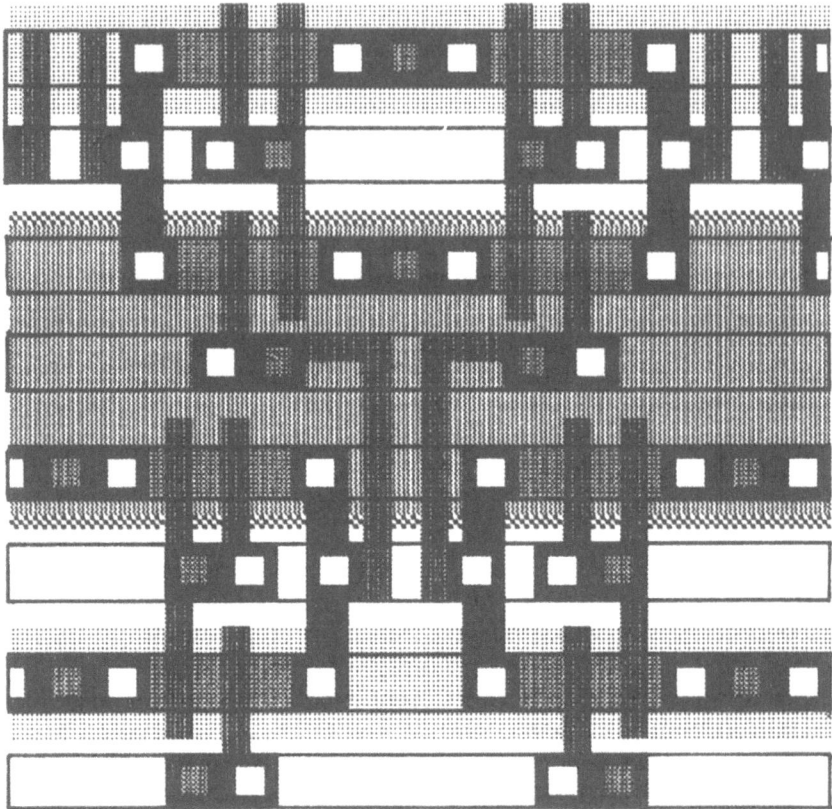

Figure 6.4 Section of a full custom 3 μm CMOS integrated circuit (schematic)

(2) Total control is exercised over the circuit's geometry. This implies that performance can be maximised, and the trade-offs between area, speed performance and power consumption finely tuned.

(3) As there is no underlying prefabricated circuitry, a full custom chip may, in principle, be no bigger than the circuit area utilised. The fabricator may, however, have a range of preferred chip frames, which will constrain the size of the chip.

(4) The compaction of circuitry possible with full custom design takes maximum advantage of the economic gains to be made by mass-production.

(5) The complexity of a full custom IC makes plagiarising its design very difficult and time-consuming. However, copying a design by-passes the enormous cost of developing and laying out the chip and a 'pirate' company can undersell the original inventor and flood the market with cheap imitations.

Many 'pirate' copies of popular ICs exist today (fifteen copies of one of the first 16K memory chips) and the United States has had to set up a

Integrated Circuits: from Concept to Silicon

Semiconductor Protection law to try to protect companies property rights in this field.

The disadvantages are of complexity, expense and manpower. They are as follows.

Disadvantages

(1) Graphical custom design implies heavy usage of computer resource and designer effort and is therefore expensive and time-consuming. There is also an 'unfortunate' phenomenon known as 'designer burn-out' which afflicts those who have been exposed to the rigours of designing custom chips. The task is full of difficulties and frustrations.
(2) A full mask set has to be made, and all steps in the fabrication cycle performed.
(3) The full custom design route is still the most error-prone, even with the use of high-quality CAD tools.

6.4 Programmable Logic Array (PLA) Design

The preceding sections of this chapter have dealt with three distinct approaches to the design of an integrated circuit. There is a further form of technique that can be used to design a gate array, standard cell or full custom IC (or sections of it) called the *Programmable Logic Array* or *PLA*. The PLA technique has enough generality, and is of sufficient importance to warrant description here, in order that its strengths and weaknesses can be appreciated. A more complete treatment can be found in Mavor *et al.* or in Weste and Eshraghian (see Bibliography at end of book).

A PLA may be regarded as simply an orderly way to organise logic circuitry. By way of an analogy, lined writing paper is a way to organise a piece of text. The text will probably take up more space than it would on a blank sheet, but it will be neater and easier to write and understand.

As an example, let us look at the function of a one-bit full adder. For three binary one-bit input numbers A, B, and C, a full adder forms Sum and Carry bits as its outputs. The Boolean forms of Sum and Carry are

$$\text{Sum} = A\bar{B}\bar{C} + \bar{A}B\bar{C} + \bar{A}\bar{B}C + ABC \tag{6.1}$$

$$\text{Carry} = AB\bar{C} + A\bar{B}C + \bar{A}BC + ABC \tag{6.2}$$

Which may be rewritten as

$$\text{Sum} = \overline{\bar{A} + B + C} + \overline{A + \bar{B} + C} + \overline{A + B + \bar{C}} + \overline{\bar{A} + \bar{B} + \bar{C}} \tag{6.3}$$

$$\text{Carry} = \overline{\bar{A} + \bar{B} + C} + \overline{\bar{A} + B + \bar{C}} + \overline{A + \bar{B} + \bar{C}} + \overline{\bar{A} + \bar{B} + \bar{C}} \tag{6.4}$$

For straightforward 'random logic' implementation of the adder function, De Morgan's theorem can be used to yield the much more concise forms

Sum = [A XOR B] XOR C (6.5)

Carry = [A XOR B] C + AB (6.6)

The adder can then be formed as three exclusive-OR gates, and one complex (OR(AND)) gate. In contrast, the PLA approach attacks equations (6.3) and (6.4) directly, using the regularity of the *minterms* (such as $\overline{A} + \overline{B} + \overline{C}$ in both equations). From this algebraic regularity, a physical regularity is imposed that simplifies the implementation on silicon at the expense of conciseness. Figure 6.5 shows the logic level concept of a PLA. Each input is assigned a physical *column* in the left-hand side of the PLA, and both true and complementary inputs (for example, A and $\overline{A}$) are directed up the column. A regular array of possible interconnect points allows a column of NOR gates to form the minterms (seven in this case). Each minterm occupies a *row* in the PLA, and the left-hand side is called the *AND plane*. An exactly analogous array of interconnects on the right of the PLA allows the minterms to be grouped to form the output signals (two in this case), each of which is assigned an output column in the right-hand side, which is called the *OR plane*. The apparent misnomers *AND plane* and *OR plane* make sense in terms of the original equations (6.1) and (6.2), not the 'De Morganned' versions in (6.3) and (6.4).

The regularity is obvious. Less obvious, perhaps, is the ease with which the PLA can have its function changed, either to form differently constituted minterms (change the interconnects marked x) or to change the function altogether (change the number of rows and columns). Figure 6.6 shows a static CMOS implementation of the section of the adder PLA shown surrounded by the dotted line in figure 6.5. The PLA is 'programmed' for its function by the inclusion of appropriate interconnect points (again marked x). In addition, transistors may be left out of the PLA where they are not required, without destroying the regularity, although this option is not illustrated in figure 6.6. Fully static CMOS PLAs are inefficient in their use of silicon area and are also slow, so a host of dynamic PLA types have been developed in addition to nMOS and pseudo-nMOS types. These are beyond the scope of this book.

A PLA is essentially an *ASIC* construct, although mask-programmable and electrically-programmable PLAs are not uncommon. The function of a PLA is usually to implement a section of an ASIC (usually random logic) quickly and accurately. It may also be used (with clocking and feedback) as a *finite state machine*. Again, we do not have space to discuss this topic in this book.

To sum up this section on PLA design, the advantages and disadvantages of the PLA are as follows.

Integrated Circuits: from Concept to Silicon 97

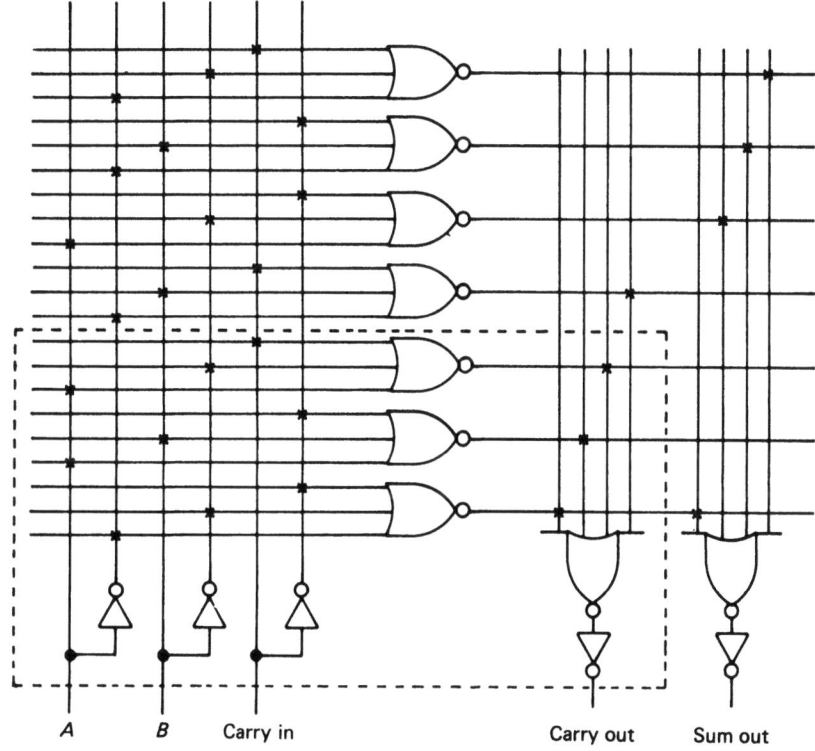

Figure 6.5 Logic level representation of a Programmable Logic Array (PLA) one-bit full adder

6.4.1 Programmable Logic Arrays: Advantages and Disadvantages

Advantages

(1) The underlying PLA structure can be pre-designed, the design rules pre-checked, and the speed performance pre-determined.
(2) The regularity ensures quick and easy PLA design.
(3) PLA design can easily be automated, and there are many 'smart' PLA generator programs.

Disadvantages

(1) A straightforward 'random logic' implementation is almost always faster and smaller and consumes less power than a PLA. However, advances in software design are reducing these problems.

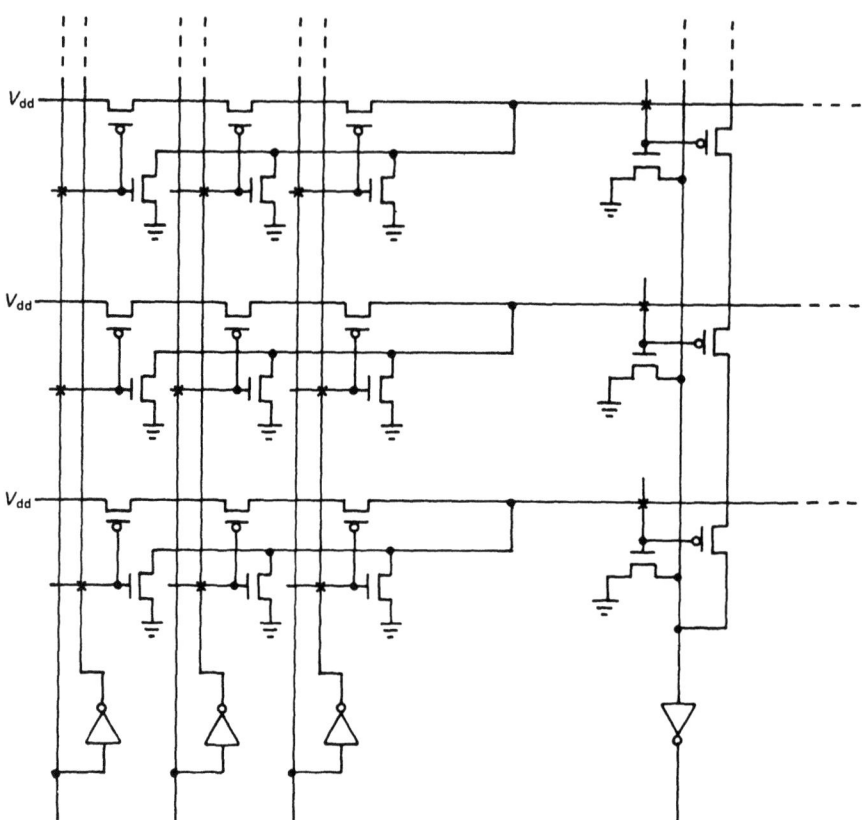

Figure 6.6 Transistor level representation of a section of the (PLA) full adder of figure 6.5, using static CMOS technology

6.5 Summary and Conclusions

The three routes from a specification to a working chip have been described. For the ultimate in performance and security, full custom IC design should be chosen. For a 'cheap and cheerful' solution, gate array design is optimal. For the enormous grey area between these two extremes either a high-performance gate array or a well-supported standard cell design route is best. In every case, for either random logic or as a state machine (particularly useful in control circuitry) the use of a PLA should be considered.

When making the choice as to which route to take, it is advisable to spend some time considering all the alternatives. The silicon fabricators offer 'deals' on performance, cost and timescales which can be substantially different, and these may make a critical difference to the final product. It may (for instance) be best to get silicon quickly and put the product on the market as soon as possible. On the other hand, if the timescale is not critical to the product's success, getting a higher performance IC may be advantageous. The likely volume of production must also be considered, as a high volume implies lower cost per chip while if a small production run is envisaged there may be problems in selecting a suitable fabricator as not all will fabricate small numbers of chips.

Often, it is the level of support and the degree of confidence which the fabricator offers that is the deciding factor. As is usual in any design and production cycle, 'you pays your money and you takes your choice'.

7 Design Discipline and Computer Aided Design (CAD) and Test

A discussion related to CAD tools may seem out of place in a book intended for electronics engineers and therefore a few words of justification are necessary. There are three major forces driving the increasing level of integration in ASICs. These are:

(1) Advances in fabrication techniques that result in smaller transistor geometries.
(2) Industry's need to put as much as possible on a single chip.
(3) Advances in CAD for ASIC design.

The importance of advances in CAD techniques in ASIC development cannot be overemphasised as, until recently, the problem of complexity management has been a severe restriction on progress towards greater levels of integration. Constant development of sophisticated Computer Aided Design (CAD) tools has become essential to enable designs comprising hundreds of thousands of devices to be designed reliably and within reasonable timescales. In addition, IC testing has become more difficult as chips have become more complex. Testing has therefore become increasingly automated and subject to discipline, and is discussed at the end of this chapter.

To give some background to the following discussion on CAD trends and their likely implications for the custom/semicustom chips of the future, we should first look at all the tasks involved in an ASIC design. The design cycle for a full custom ASIC is shown in figure 7.1. There are two fundamental activities: *design proper*, which proceeds in a downward direction as modules are successively decomposed in figure 7.1, and *verification*, which proceeds upwards. At every stage in the design, more detail is added to define exactly *how* the various modules required are constructed. *At the time they are added*, these details must be verified or checked for correctness against both the ASIC specification and against any previous attempt at the module. The verification process consists, therefore, of a two-way interaction in which module descriptions are scrutinised

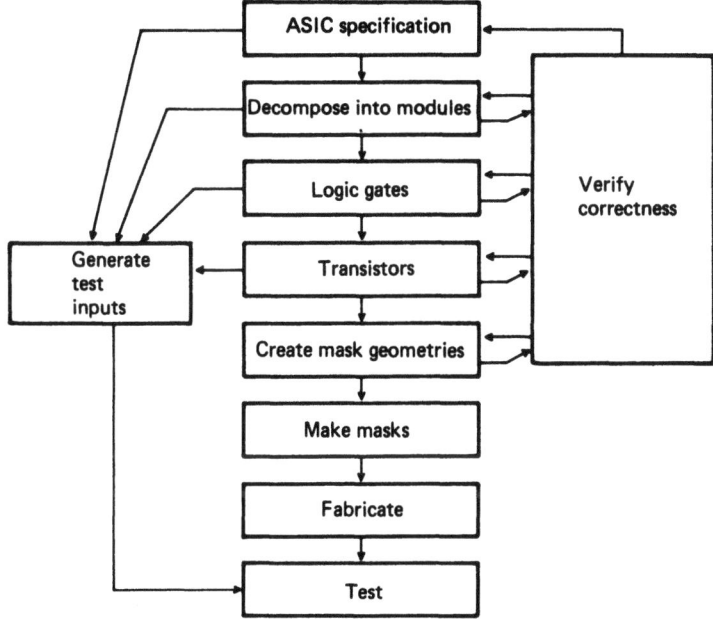

Figure 7.1 Design cycle for an integrated circuit

for either consistency with a previous less-detailed version or correctness with respect to the ASIC specification.

All the procedures involved in ASIC design and verification are error-prone, and any automation that increases speed, integrity and ease of design is advantageous. The need for verification is now met by a vast range of *simulation* programs, which enable the designer to build a software model of a circuit (an entire ASIC, a module, submodule, logic gate or transistor). The simulator will then mimic the circuit and predict its behaviour (to within the accuracy of its model).

Unfortunately, CAD programs are not as good as human beings at the intuitive thinking that creates sections of dense ASIC layout, and so automation of the design process has been less complete. Currently, if *design* automation is used, some area inefficiency results and it is important to be clear about the extent to which this can be tolerated when the advantages of automation are taken into account.

Before a chip can be either used or sold it must be tested thoroughly to detect failure owing to imperfect fabrication, bonding or packaging. The generation of suitable input signals which will put the chip through its paces, a topic well outwith the scope of this text, is well supported by CAD tools but may still require some intuition and help from the ASIC designer.

The following sections describe trends in ASIC CAD tool development, but no claim is laid as to the treatment being either definitive or complete. Before plunging into the realms of CAD, however, we must examine the underlying principles of good ASIC design, which are based on the concepts of hierarchical design techniques and the use of *macros*.

7.1 Hierarchical Design and Macros

The most important single concept in good ASIC chip design is that of hierarchical design. This intimidating phrase simply describes what good computer programmers have been doing for years, and illustrates the way in which the disciplines of hardware and software design are influencing each other. By writing subroutines or procedures that are called by a main program (and by each other) to carry out a given computational function, programmers are able to generate compact, efficient, well-structured and understandable programs. There is nothing worse (and more likely to go wrong) than an amorphous mess of program lines in which procedures/subroutines are never used. To translate this concept into the language of chip design, subroutines are replaced by circuit modules, which may be as small as a NOR gate or as large as a CPU. In this book these modules will be called *macros*, though they can also be referred to as 'cells' or 'groups'. The reasons for using hierarchical methods in ASIC design are similar to the reasons for writing hierarchical (structured) programs. They are:

(1) The growing complexity of ASIC design can only be managed when good use is made of hierarchical design, as otherwise designs become unbearably cumbersome. It is inefficient to store, transfer and manipulate unnecessary, redundant data by describing separate versions of the same circuit element repeatedly. To show the scale of the problem, consider a typical chip consisting of 500 000 transistors. For the transistors alone, there will be $3 \times 4 \times 500\,000$ separate values required, since each transistor consists of at least three rectangular shapes, each defined by two pairs of coordinates. If each coordinate occupies 2 bytes, this results in 12 Mb (12×10^6 bytes) of data. When the interconnect data is included this can be trebled, giving a final data file size of approximately 36 Mb. Attempting to manipulate this amount of data will bring most computers (and ASIC designers) to their knees!

(2) If, for example, 20 NOR gates are required in a circuit, it is absurd to design 20 different NOR gates. It is better to put effort into designing one good NOR gate, and placing it (or 'instantiating' it) at the desired 20 locations. This minimises the opportunity of making errors.

(3) If an error is discovered in a gate and that gate has been defined 20 separate times, each of the definitions will have to be modified. If, however, *one* gate has been defined and stored and has then been called 20 times, only the

master copy need be changed. This will cause *all 20 instances* to be updated accordingly.

Macros or layout subroutines for patterns that occur many times within a design are therefore a most important feature of any CAD package. Indeed, the availability of a library of standard macros for commonly used functional elements such as logic gates and flip-flops is a feature of every commercial gate array design system. Figure 7.2 shows a series of 'nested' macro calls schematically. The macro NOR is called by the larger macro MUX (for MUltipleX), which is in turn called by a macro called MULTIPLIER. The MULTIPLIER macro forms part of the main chip definition. Note that NOR is also called by the CONTROLLER macro, and any change made to NOR will affect each of CONTROLLER, MUX and MULTIPLIER.

7.1.1 Top-down Design and Partitioning

Allied to, and implicit in, the discipline of hierarchical design is the concept of 'top-down' design and partitioning. In a 'top-down' design the chip's intended function is specified (at the 'top' level) by a truth table, or set of input/output patterns that do not show the internal workings of the chip. This 'top-level' function is then divided into subproblems, or modules, which may themselves be further subdivided many times.

For example, in the GATEWAY exercise described in section II, one of the 'top-level' problem specifications can be given as 'add two numbers together then compare them with a set of upper and lower limits'. Obviously, this prob-

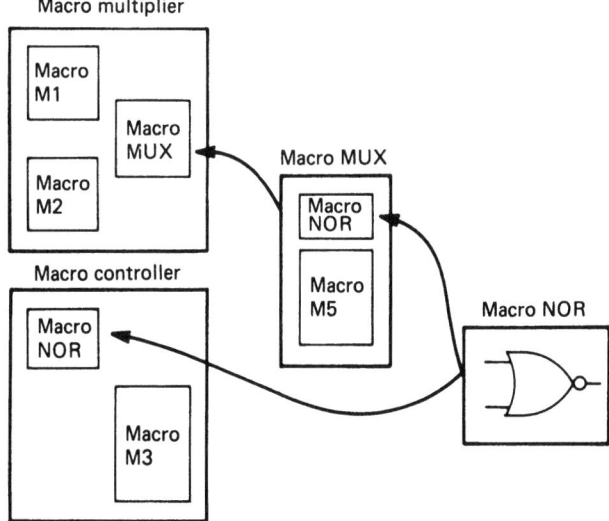

Figure 7.2 Use of circuit subroutines or *macros*

lem can be split into two modules – 'add' and 'compare'. It would not be appropriate here to give many more clues to how the GATEWAY exercise should be partitioned, but it is also obvious that a subpartitioning of the 'compare' operation into 'compare with upper limit', and 'compare with lower limit' will get us a little closer to a good design. This process of partitioning modules repeatedly continues until the descriptions of the latest 'bottom level' modules can be realised as a network of logic gates and MOS transistors. The logic design is then complete, and the detailed interconnect pattern may be developed as the next stage.

If we design a chip interconnect pattern using only a 'top-down' approach, without having first designed the low level macros, we find ourselves in a situation analogous to designing a complete house without knowing if assumptions made about the sizes and properties of basic components, such as bricks, windows and doors, are valid. Clearly, this would result in a peculiar house!

If we tried designing our chip interconnect using only a 'bottom-up' approach without considering the overall layout of the chip (the 'floor-plan') we would eventually find ourselves in a situation analogous to having designed our house only to find that it will not fit on the plot of land available!

We must use a combination of 'top-down' and 'bottom-up' design techniques. We need the 'top-down' in the GATEWAY exercise to ensure that the chip 'floor-plan' is correct so that efficient use is made of the array cells. In a custom design we must carry out the same sort of operation to ensure that the final chip will make an efficient use of silicon area and will be of an appropriate size and shape. However, detailed design must be carried out from the 'bottom-up', using the same hierarchy of macros developed to represent the logic functions. Thus the fundamental gates (NAND, NOR, etc.) must be designed first.

To begin with, however, it is worth describing in more detail the function and scope of simulation CAD tools, as these should be used extensively *before this process of chip layout begins.*

7.2 Design Verification by Simulation

Each time the designer partitions a problem into modules, he can make mistakes. He will be using some combination of truth tables, Karnaugh maps and standard logic minimisation techniques (according to his preferences) to ensure the correctness of each partition. However, no matter how careful he is, mistakes can occur, and it is desirable to have a second opinion as to the correctness of his work. Such a second opinion can be obtained from a computer simulation which will corroborate (or otherwise) the results of his Boolean Algebra. The principle of logic simulation is an extremely important one in many disciplines, but nowhere more than in chip design.

A chip design is made up of a number of elements: logic blocks, logic gates and, ultimately, the MOS transistors themselves. A logic simulator is a computer program that mimics the behaviour of these elements in the design. To test the logic, different inputs are fed to the program and the resulting outputs are examined for correctness (as one would test the completed chip on an oscilloscope or logic analyser). The simulation will, however, only be as good as the computer's model of the components. This implies that, while a behavioural simulator (which models logic functionality via programming subroutines) is ideal for initial logic development, a more detailed (switch level) model of transistor behaviour is required for later circuit validation. The following sections describe the different types of simulation and simulators used in an ASIC design.

7.2.1 High Level Simulation

If you, as a circuit designer, check the correctness of your circuit schematic by following voltages (either logic levels or actual voltages) manually through your circuit, you are doing a simulation. CAD simulators for circuit design, which automate this tedious task, have been used for some time by system and circuit board designers, and many simulators are aimed more at the board designer than at the ASIC designer. However, most of the current generation of simulation tools have Integrated Circuit design firmly in mind, and many are written specifically to support this activity. There are many ways, involving different degrees of accuracy, of simulating a section of ASIC circuitry. To understand this concept, imagine checking the correctness of an ASIC schematic diagram by hand. First of all, the designer checks that the overall block structure of his ASIC is correct. He then checks that the internal structure of each block, represented as networks of logic gates, is correct by stepping logic values through the circuit. In addition, the designer may wish to be particularly careful about some critical sections of the design, and may spend some time checking voltages in the relevant transistor networks to convince himself that they are correct. This activity is a form of simulation at a set of different levels of detail. At each level of extra detail, the designer is gaining more information about, and therefore confidence in, his circuit.

We can illustrate this by an example. If as part of an ASIC, a multiplication function is to be realised, the designer has to design and verify a block of layout that has as its inputs two numbers A and B, and as its output a third number C. The simplest form of computer simulation that can be performed is to write a simple program (in pseudo-BASIC) to simulate a multiplication function, such as:

```
10   INPUT A
20   INPUT B
30   C = A*B
40   PRINT C
50   STOP
```

This form of simulation is fast and easy to understand, but gives no information whatsoever about the correctness of the section of multiplier circuitry.

If, however, the designer has decided that multiplication consists of a series of SHIFT and ADD operations, the little program just given may be expanded to give a more realistic simulation involving more detail. Eventually, the designer will generate a version of his multiplier circuit that takes the form of an interconnected set of logic gates and registers (latches) on a schematic diagram. The next step is to do a *gate level simulation*.

7.2.2 Gate Level Simulation

Figure 7.3 shows a small section of a multiplier. The circuit shown implements the Boolean functions $D = (A \text{ XOR } B) \text{ XOR } C$ and $E = (A \text{ XOR } B)C \text{ OR } AB$. The input and output wires, or 'nodes', have been labelled as in the equations. In most logic expressions of any complexity, intermediate logic values are calculated. These are 'internal nodes' in a logic schematic, such as the output of the upper XOR gate and the two AND gates in figure 7.3. They have been given (arbitrarily) the names F, G and H. A useful analogy here is that of the local variable in a software subroutine or procedure, because such a variable is 'forgotten' by the program as soon as the subroutine has completed its function. The *Gate Level Logic Simulator* is the most popular and available type of circuit simulator and has the capability of modelling circuitry at this level of detail. The circuit to be simulated is described in a particular Hardware Description Language (HDL). The section of multiplier circuitry shown in figure 7.3 might be described in the form of an HDL for input to a Gate Level simulator as follows:

```
XOR_1 =    XOR_GATE (A,B,F)
XOR_2 =    XOR_GATE (F,C,E)
AND_1 =    AND_GATE (A,B,G)
AND_2 =    AND_GATE (F,C,H)
OR_1 =     OR_GATE (G,H,D)
```

A section of circuit description which is defined in the same way as a software subroutine (or procedure), and is then 'called' by another circuit subroutine (or the main ASIC definition) is called a 'macro'. Each reference to a macro is called an 'instance'. The two instances of the XOR_GATE function have to be given unique names (XOR_1 and XOR_2) to distinguish them from one another. The syntax of this (fictitious) simulator HDL is such that 'XOR_1 = XOR_GATE(A,B,F)' says that XOR_1 is an exclusive-OR gate (XOR_GATE) macro and is called such that the inputs are nodes A and B, and the output is node F. This means that $F = A \text{ XOR } B$. The section of circuitry described by this piece of HDL may form the complete circuit to be simulated, or may itself be a macro called from elsewhere.

Typically, a gate level simulator, while accepting and delivering only logical '1' and '0' levels during the simulation, will be able to model 'unknown' logic states. These occur, for instance, when the outputs of two logic gates are con-

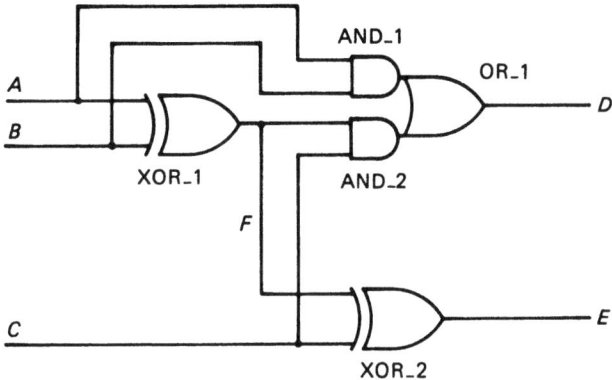

Figure 7.3 Section of an integrated multiplier (schematic)

nected to a single circuit node. If the two gates disagree about the logic value on the node, the output of the simulator will say that the node in question is in an unknown state. The simulator may also understand and deal with 'high impedance' and 'dynamically stored' logic values (when the presence or absence of charge on a node defines its logical value). The details of such modelling are outwith the scope of this book. In some simulators, results may be shown in graphical form, looking like the output of an oscilloscope. More often they look like the truth table shown below.

Typical (truth table) form of logic gate level simulation output

A	B	C	F	G	H	D	E
0	0	0	0	0	0	0	0
0	0	1	0	0	0	0	1
0	1	0	1	0	0	0	1
0	1	1	1	0	1	1	0
1	0	0	1	0	0	0	1
1	0	1	1	0	1	1	0
1	1	0	0	1	0	1	0
1	1	1	0	1	0	1	1

The values of A, B and C are inputs to the section of circuit under design, and are therefore defined by the designer to test his circuit model. The values of $D-H$ are calculated by the simulator, and they are the simulation results. Such a simulator is of enormous value because, by performing a simulation such as in this simple example, the designer can study the output of his circuit for a given set of inputs much more quickly and accurately than he could ever hope to do manually. Also, during a manual simulation it is easy to become convinced that a circuit is correct, particularly if the same person who designed the circuit is checking it. The simulator can act as an impartial friend who does not make such mistakes! Gate level simulators are likely to have useful extra capabilities such as the ability to model time delays (crudely), and to take parasitic capacitances and the associated degraded rise times and fall times into account. It may also be possible to use such a simulator to model circuits with faults inserted (valuable when checking the suitability of proposed test patterns).

7.2.3 Switch Level Simulation

Once a section of ASIC has been verified (that is, simulation has shown that a gate level description of the circuit is correct) it is desirable to perform a 'switch level simulation'. Switch level simulation allows the modelling of MOS transistors as switches, and logic gates (or higher level macros) as networks of transistor switches. This takes the simplifications made in chapter 2 one stage further by treating transistors as pure bidirectional logic switches. The result of a switch level simulation of the circuit in figure 7.3 should be identical to that shown above (otherwise the implementation of the logic gates as transistors is wrong!). It is not appropriate to go into much more detail here, but it is worth presenting a typical circuit description for a switch level simulator (again imaginary).

Representation of a circuit for the Boolean exclusive or function in the hardware description language of a switch level simulator.

```
START   XOR_GATE(A,B,C)
N1 =    NTYPE (g = A, d = C, s = G)
N2 =    NTYPE (g = B, d = G, s = GND)
N3 =    NTYPE (g = D, d = C, s = GND)
N4 =    NTYPE (g = B, d = D, s = GND)
N5 =    NTYPE (g = A, d = D, s = GND)
P1 =    PTYPE (g = A, d = F, s = V_DD)
P2 =    PTYPE (g = B, d = F, s = V_DD)
P3 =    PTYPE (g = D, d = C, s = F)
P4 =    PTYPE (g = B, d = D, s = E)
P5 =    PTYPE (g = A, d = E, s = V_DD)
END     XOR_GATE
```

The circuit described by this piece of simulator code is shown in figure 7.4, and is the CMOS XOR gate of chapter 2. The node (or wire) names are *not* the same as those in figure 7.3 and the associated piece of gate level code, as the description is appropriate to a different simulator with a different description language syntax. They are internal names, similar to local variables in a subroutine, and the description casts the XOR_GATE as a circuit macro (enclosed by 'START' and 'END' statements). As before, individual instances of the NTYPE and PTYPE transistors must be labelled uniquely to avoid confusion (for example,

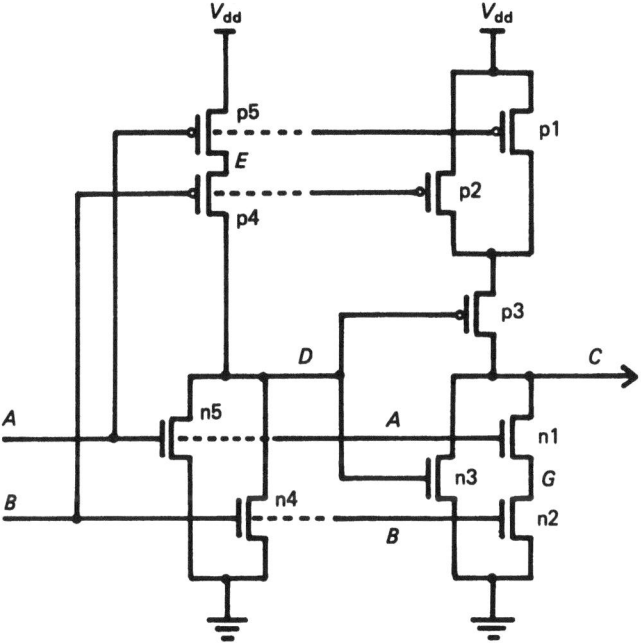

Figure 7.4 A CMOS exclusive-OR (XOR) gate

N1, N2, etc.). Also, the gate (g), source (s) and drain (d) connections must be specified and have associated with them internal node names (*D-F*) as necessary.

Once a circuit subroutine has been described in this way, it can be called up as often as necessary to form the complete circuit. The following is the hardware description language for a circuit instancing the switch-level exclusive-OR gate defined above.

START MAIN_CIRCUIT
XOR_1 = XOR_GATE (*A* = *A*, *B* = *B*, *C* = *F*)
XOR_2 = XOR_GATE (*A* = *F*, *B* = *C*, *C* = *E*)
etc., etc.
END MAIN_CIRCUIT

Once again, the two instances of the XOR_GATE macro must be uniquely named (XOR_1 and XOR_2). The perverse-looking assignments in the brackets define the correspondence between the node names in the *description* of the XOR_GATE (as given in figure 7.4) and those in the *instance* of the XOR_GATE (as given in figure 7.3). For instance, the bracketed expression in the instance of XOR_2 implies that nodes *A*, *B* and *C* in the *description* of the XOR_GATE macro are respectively nodes *F*, *C* and *E* in this *instance* (XOR_2) of the XOR_GATE.

A switch level simulator operates in fundamentally the same way as the gate level simulator in that all its signals are either logic '1' or logic '0', but it provides more detail and therefore more simulation accuracy. It is of little use to a *systems* designer, who will not be dealing with individual transistors, but is of enormous value to the ASIC designer, who needs a switch level simulator to verify that his transistor level implementation of his ASIC function is *logically* correct.

7.2.4 Device Level Simulation

A thorough examination of this topic would go far beyond the bounds of this book but we should state what is involved. A *device level simulator* uses approximations to the full device equations given in chapter 2 to model the *analogue* behaviour of a circuit. In other words, transistors are treated as transistors (not just switches) and voltages are treated as voltages (not just logic '1' and logic '0' levels). This form of simulation is expensive in CPU time (and therefore in money!), and is also time-consuming. It is sometimes omitted (which is dangerous) and is often confined to only those sections of ASIC circuit whose performance is of major importance. Certainly, a device level simulation of an entire ASIC is an outrageous proposition.

This has been a lightning tour of simulation. There are many interesting areas (fault simulation, hardware simulation acceleration, etc.) to which we cannot devote time here, and advanced books have already been written on the subject. To conclude: the ASIC designer needs a range of simulation tools to support the verification of his circuit's correctness at different stages in its development. When verification via simulation is omitted at any stage, all work from that point onwards is at risk from undetected errors and the cost in wasted time and effort can be disastrous.

7.3 Physical Design and Chip Layout

When the 'top-down' design has been completed and a transistor network has been derived, simulated and found to be correct, an interconnect pattern representing this network can be created. This work could be carried out using textual representations of tracks in Cartesian coordinates, but it would be tedious and difficult to check. It is much better to use graphical CAD tools to 'draw' metallisation tracks on a computer screen, or even to automate the entire process.

There are more CAD routes from a circuit schematic to silicon than the three (Gate Array, Standard Cell and Full Custom) hardware routes described in the previous chapter. It is only possible to describe a selection of them here to give an impression of the way in which automation is progressing. This will be dealt with by considering first the *least* automatic method, and looking subsequently at progressively more automated methods.

7.3.1 Traditional Graphic Editors

These CAD systems, which give the designer the ability to design shapes on a computer terminal (rectangles, polygons, etc.) were once ubiquitous. Now their role in the design of complex ASICs is reduced to the creation of either sections of analogue circuitry, the design of the *contents* of a standard cell library, the design of the *underlying structure* of a gate array, or the design of 'oddball' bits of circuitry (for example, memory) that are not amenable to more automated methods. The section of full custom nMOS circuitry shown in figure 4.6 was designed using a colour graphic editor. Creating designs in this way offers the greatest control over the ASIC's performance, and is also the most instructive from the educational point of view.

Currently available graphic editors offer the ability to perform 'design rule checks' on sections of circuitry, and some even check every shape as it is created. Designing entire ASIC circuit elements in this way is certainly possible (until relatively recently it was the only technique available) but it can be rather painful.

7.3.2 Symbolic Editors

This technique has several manifestations that all make use of hierarchical design techniques and recognise, for instance, transistors as single (often parameterised) symbols. For instance, a 'symbol' may be placed that corresponds to a p-type enhancement device of width 7 μm and length 3 μm. This action will place all the shapes that form the transistor (active area, polysilicon, P-implant, etc.) correctly and simultaneously. Symbolic editing facilitates on-line design rule checking, as much of the complexity of inter-shape distance checking is taken care of by the use of symbols. Recently, the use of Artificial Intelligence techniques have been shown to be advantageous at the symbolic level, with the design system performing rule-based automatic layout and routing, but the full promise of this work is yet to be realised in silicon.

7.3.3 Automatic Routing and Placement

This form of design automation is relevant to standard cell and gate array ASIC design and also to full custom ASIC design where large sections of layout created by a graphic editor have to be joined together. For some time CAD systems have been available that place large 'black box' sections of circuit and subsequently generate the interconnect wiring, or 'routing', automatically. The efficiency of such systems depends critically on such factors as good system partitioning, the shapes of the modules to be placed and interconnected, the number of interconnect layers available and the choice of a 'floor-plan' style for chip organisation.

The sections of circuitry shown in figures 5.1, 5.2 and 5.3 could have been created by an automatic place-and-route program. As was mentioned in chapter 5, a formal style for the basic floor-plan makes it easier to automate the routing and floor-planning. 'Streets-and-houses' gate arrays are therefore easier to auto-route than 'sea-of-gates' arrays. For the same reasons, standard cell libraries with regularly shaped cells are easier to handle automatically than those with cells of all shapes and sizes. In general, an automatic place-and-route design system will require a circuit description in a form similar to that given above as an example of a logic level circuit description (figure 7.3 and associated simulator code). The form of this Hardware Description Language (HDL) differs between autorouting systems, and some offer the option of graphical capture of a schematic diagram.

Any decent auto-place-and-route system will allow critical wire lengths to be controlled. A good one will, at the end of placement and routing, reconstruct the circuit description with the parasitic capacitance values added so that the circuit can be re-simulated. This is dealt with in a separate section below.

7.3.4 Silicon Compilers

Most contentious, and to some sceptics merely a piece of marketing jargon, are Silicon Compilers. Strictly speaking, a Silicon Compiler accepts a description of an ASIC function in a high level or even schematic form, with the hierarchy of functions represented in a form analogous to procedural hierarchies in high level language computer programs. From this input, coupled with some information about what outputs the ASIC should produce under known stimulation, a silicon compiler will simulate the ASIC function and also generate the entire ASIC layout with little or no human intervention. It will also produce simulation results and mask data. It is the similarity between these products and the diagnostics and machine code produced by a software compiler that has led to the term 'Silicon Compiler'.

A Silicon Compiler calls on pre-defined and pre-verified circuit blocks and assembly techniques, so circuit integrity is guaranteed once the compiler is fully debugged. In common with all other automatic layout systems, a Silicon Compiler has the disadvantage that the circuits it produces take up more silicon area than they would if they were produced as hand-packed designs. Clearly, the increase may sometimes be intolerable, and this, coupled with the restricted applicability of individual silicon compilers, discourages many designers from using them. Silicon Compilation is also a heavy user of CPU time, but as increased computing power is offered by advanced (Silicon Compiled ?) chips, this should not be a long-term worry. The advantage of a Silicon Compiler is that it can allow designers to carry out in a few days a complex ASIC design which is guaranteed correct, and this advantage is so great that these tools are bound to have a bright future.

7.3.5 Circuit Extraction and Re-Simulation

Once the designer is satisfied that his interconnect pattern is complete and correct, he may feel ready to have the device fabricated. In a manual chip layout, however, errors may have been made in the translation of his transistor network to an interconnect pattern, while in an automated chip layout the extra parasitic capacitances due to the interconnect wiring may have cast doubt on the validity of earlier simulation results. It is therefore valuable to check the design at this stage by running another computer simulation which takes as its input the *actual* transistor interconnections. This should show how the final circuit will perform, and point out any differences between the chip's intended function and its actual function.

Automatic layout systems should generate a transistor interconnect description annotated with parasitic capacitances. This can be simulated to confirm that the circuit still conforms to the specification. In a full custom (handcrafted) design, a *circuit extractor* must be used to generate, from the mask data, the transistor interconnect pattern and parasitic capacitances, in order that re-simulation may be performed. This is a massive task which is expensive in terms of engineering time and computer resources, but the cost of a failed design is also high and checking at this stage is almost essential.

7.3.6 Generation of Fabrication Data

This is neither an interesting nor an educational task, although it absorbs a lot of computer resource. The hierarchy of nested macros, etc. which represents the interconnect pattern must be 'smashed out' to a long list of line segments, and subsequently translated to a list of the rectangular shapes that will be used to make the photographic mask. Often it is only incumbent on the designer to enure that the correct database file is made available when this operation is carried out, but some IC fabrication companies also require a variety of different types of extra information. Only a very inexperienced designer will not check these details before be submits his design for fabrication.

7.4 Integrated Circuit Testing

IC testing is a wide-ranging subject, and it is impossible to give more than a brief description of the topic in this book. The Bibliography (at the end of the book) shows where more detail may be found. This section attempts to give the flavour of IC test, however, and indicates the problems and solutions that occur.

7.4.1 Test Pattern Generation

A test system is a generator of input patterns that will, to some degree, exercise the nodes of the chip to be tested. In other words, the output from the chip undergoing testing, under known and controlled stimuli, is compared with that for a circuit that is known to be functional, or with the results of simulation. If these outputs are different the device under test is deemed to have failed, and it is discarded. All test strategies rely on this general method. The major decision that has to be made is the degree of rigorousness of the test, and therefore the length of the set of test inputs. This decision will vary, depending on the circumstances. For instance, an IC for a £1.99 digital watch does not have to be rigorously tested, as the failure of such a watch is merely a nuisance. An IC that forms part of a life-support system or a missile clearly does, as its failure would be less acceptable. We should therefore look at the different methods of arriving at a set of test inputs. These methods have been subjected to some degree of automation, and the resultant CAD software is classed as Automatic Test Pattern Generation (ATPG). Many ATPG programs will derive a minimised set of test patterns from a netlist (as described above) of the circuit to be tested.

Exhaustive test

Circuits can be tested by using an *exhaustive* selection of all the possible binary input patterns but for practical circuits this type of test usually takes too long to complete. This is particularly true if the circuit contains sequential logic, as then every possible set of *internal* states must also be probed. Table 7.1 shows an exhaustive test for the XOR gate of figure 7.4.

Table 7.1 Exhaustive Test of XOR Gate. (1) = Stuck Fault

A	B	D	E	F	G	C
0	0	1	1	1	X	0
0	1	0	1	1	0	1
1	1	0	X	0(1)	0	0(1)
1	0	0	0	1	1	1

Application of all four input patterns (00, 01, 10 and 11) constitutes an exhaustive test. This tests for all possible short-circuits between circuit nodes and the 0 V and 5 V supply lines. These are termed 'stuck at 0' and 'stuck at 1' faults respectively, and one such fault (node *F* shorted to 5 V) and its consequences are shown as bracketed values. Most test pattern generation strategies rely on testing for 'stuck at' faults, although many other forms of fault (for example, inter-node short-circuits) exist. The adoption of 'stuck faults' as a *fault model* allows an impressive simplification of the test procedure to be made, although it is not physically realistic. In spite of the unrealistic nature of the stuck fault model, its use has not so far resulted in large numbers of faulty chips passing tests based on stuck fault detection. In the example given, the fault will be detected on the third test cycle (AB = 11), as an erroneous output will then be produced. This form of test reduction is invariably used in practice to avoid applying unnecessary test vectors. Exhaustive test patterns are easy to derive, and offer high test *quality*, at the expense of long test times. To maximise the usage of the (extremely expensive) test equipment it is essential that testing is as rapid as possible, while maintaining test quality.

Structural and functional test

Let us take as an example the chip in figure 7.2 that consists of a multiplier, a controller and a NOR gate. It should be possible to apply a set of test vectors that put the multiplier through its paces while disabling the controller, followed by a second set that tests the controller and disables the multiplier. This strategy clearly exploits knowledge of the device's internal structure, and is therefore a *structural* test. It offers a substantial reduction in test length over an exhaustive test. For instance, if a 4-bit multiplier has 256 possible sets of inputs and the controller has 128, an exhaustive test will consist of $256 * 128 = 32\,768$ patterns, while a structural test will only require $256 + 128 = 384$. In other words, the exhaustive test will take *85 times longer* than the structural test. Structural test patterns are more difficult to derive, but offer an excellent compromise between high test quality and short test time.

A *functional* test is easier to visualise. For the chip in figure 7.2, a crude functional test will check that the multiplier will multiply numbers, and that the controller will produce control signals. The chip can therefore be said to *function* correctly, but no attempt has been made to test for specific faults. Functional tests are generally easier to generate, shorter to apply, and the least accurate.

7.4.2 Test Pattern Validation

Once a set of test patterns has been derived, either manually or by ATPG software, it should be checked for the level of 'fault coverage' it offers. The method used for this validation *of the test patterns themselves* is fault simulation, and it is often performed by a simulator such as has been described in the preceding sections in this chapter. For a simple example, we can look at table 7.1. The truth table shown could have been produced by:

- Running a simulation of the XOR gate.
- Running a simulation of the XOR gate with node F clamped at logic '1'.

This fault simulation shows that a test consisting of only the first three rows of table 7.1 exposes this particular fault. In practice, a simulation will also be run with 'stuck at 0' and 'stuck at 1' faults on *every* node, to check that each of these faults is 'covered'. This is a lengthy procedure!

7.4.3 Design for Testability

Care must be taken during the design phase to ensure that circuits are made easily testable by, for example, providing access to internal nodes. This enables feedback loops to be broken and memory elements to be initialised. There are many other rules that apply to different types of circuit. Each designer (or company) will have its own favourite set of techniques for ensuring testable designs. Ignoring these rules incurs the danger of designing chips that cannot be tested, which does not endear the designer to the quality control department!

7.4.4 Application of Test Patterns

In this short discussion on testing it is only possible to say that applying the required test patterns to the device under test is difficult, expensive and time-consuming. The level of difficulty is increased when high speed circuits are involved, as even higher speed test equipment is required. For simple circuits, test equipment can be as simple as a signal generator and an oscilloscope, but more usually the necessary equipment costs hundreds of thousands of pounds.

In summarising this section on CAD, it is worth remarking that much of the inevitable jargon (some of which we have been unable to avoid) generated by CAD wizards merely describes what good engineers have been doing for years! In effect, this is what advances in CAD are aimed at achieving, and the results are bound to have dramatic effects on design timescales and complexities.

Section II

8 The GATEWAY Gate Array Design Exercise

8.1 Introduction

GATEWAY is an exercise in Integrated Circuit design that is aimed at pre-final year honours students. The exercise, which is intended to extend over the entire academic year, gives participants a wide variety of experience in the execution of a complete ASIC design cycle. Integrated Circuit design, technology and testing, together with Computer Aided Design (CAD) usage, project management and documentation are all valuable educational areas covered.

The exercise has run successfully since 1982 at several establishments, using textual Computer Aided Design (CAD) design software on a mainframe computer. Using cheap, powerful home computers with strong graphics capabilities, GATEWAY has now been transformed to a more industrially realistic form based on a graphic editor for design creation, running on a microcomputer workstation. This exposes students to a facet of design, graphical CAD, which has previously been unthinkable in all but limited final year project work, since its cost has been too high for educational establishments. Working on many graphical data entry workstations as opposed to textual input to a remote mainframe computer, students can perform more and better designs. The use of personal computers in CAD/CAE is increasing worldwide and the new concept is in keeping with this trend. There is little point in using expensive hardware for tasks where simple hardware, carefully programmed, will suffice.

8.2 The GATEWAY Philosophy

Central to the success of the GATEWAY concept is the constraint that every aspect of the project definition, technology and CAD tools is targeted firmly at maximising the *educational* value of the tasks involved. This constraint does not always result in an exact concurrence between detailed industrial practice and

120   *Integrated Circuit Design*

the approach adopted here. For instance, in commercial gate array design, placement and routing is performed automatically by commercial software running on a minicomputer/workstation. This results in the fast and efficient implementation of gate arrays with high design integrity, but is educationally of little value. The use of such a system for an undergraduate exercise would certainly offer training in logic design and typing, but would fail to instil any significant knowledge of the problems of physical layout and Integrated Circuit technology. Furthermore, the cost and complexity of such systems precludes their use in all but final year honours work. The GATEWAY exercise resembles full custom Integrated Circuit design, where interactive graphics CAD tools are available to carry out the interconnect. In this way, CAD experience is increased, and the problems of gate array routing are encountered directly. Even if the student never touches a graphics editor again and works exclusively with automated layout systems, he will appreciate the issues involved in selecting a floor-plan and routing strategy, and will be able to assess the strengths of different routing algorithms. The exercise can be performed as a 'paper' design without the aid of the CAD tools.

The 'tried-and-tested' GATEWAY exercise is based around a 5 μm nMOS Gate Array. As most current Integrated Circuits use CMOS technology, a CMOS version of the GATEWAY exercise has been produced. In order that the reader may choose which technology to study via GATEWAY, portions of section II of this book have been replicated for nMOS (section 8.3) and CMOS (section 8.4). The remainder is technology-independent except where noted, and serves for both the CMOS and the nMOS exercise.

8.3 The ED500 Gate Array

The gate array to be used is an n-channel silicon gate (nMOS) design (device number ED500) containing a total of 552 logic cells (24 x 23). A plot of the device is given in figure 8.1. The logical performance of the chip is determined by how these cells are configured and interconnected. The ED500 also contains 40 peripheral cells which can be configured as either input or output bonding pads. This gate array has been developed specifically for the GATEWAY exercise in the Department of Electrical Engineering at the University of Edinburgh.

Each ED500 chip will support two separate designs with common data inputs. Figure 8.2 highlights the two working areas on the array, one of which is assigned to each of the designers. Figure 8.2 also shows how the data inputs and outputs are pre-defined and wired to points near the edges of the working areas.

The ED500 device is fabricated on a process which has a minimum feature size of 6 μm. The single cell of the device, shown in figure 8.3a, contains a total of four MOS transistors. Figure 8.3a also shows some pre-defined interconnections or underpasses within the cell. These interconnections, which may be thought of as 'subways', exist deep within the gate array structure as sections of

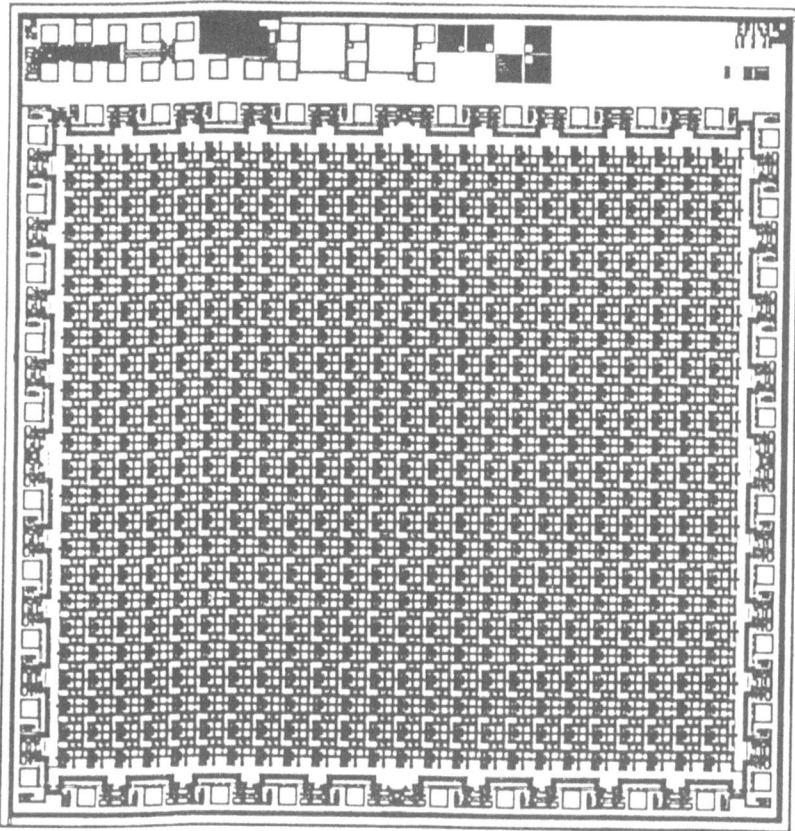

Figure 8.1 The ED500 Gate Array

diffusion or polysilicon and do not impede any metal tracks ('highways') which cross over them. The power supply rails (0V and 5V) exist as two horizontal tracks at the top and bottom of the cell.

In figure 8.4a the basic ED500 gate array cell has been redrawn in schematic form with an 18 μm grid to identify the various features to which connections can be made. The grid is labelled with indices '0' through '9' in the x-direction and indices '0' through '8' in the y-direction. This coordinate system will be used throughout the following discussions. Note that all connections must be made by paraxial lines (that is, along the dotted grid lines) and that *no angled lines can be permitted with the ED500*. This restriction is common in ASIC design, as it simplifies automated layout, and it does not sifnificantly impede layout efficiency. (The one area of ASIC design where non-orthogonal lines are *de rigueur* is in memory design.)

Estimates of likely chip performance must take account of the *lateral underdiffusion*. For the process used, this is 1.25 μm so the *actual length* of each

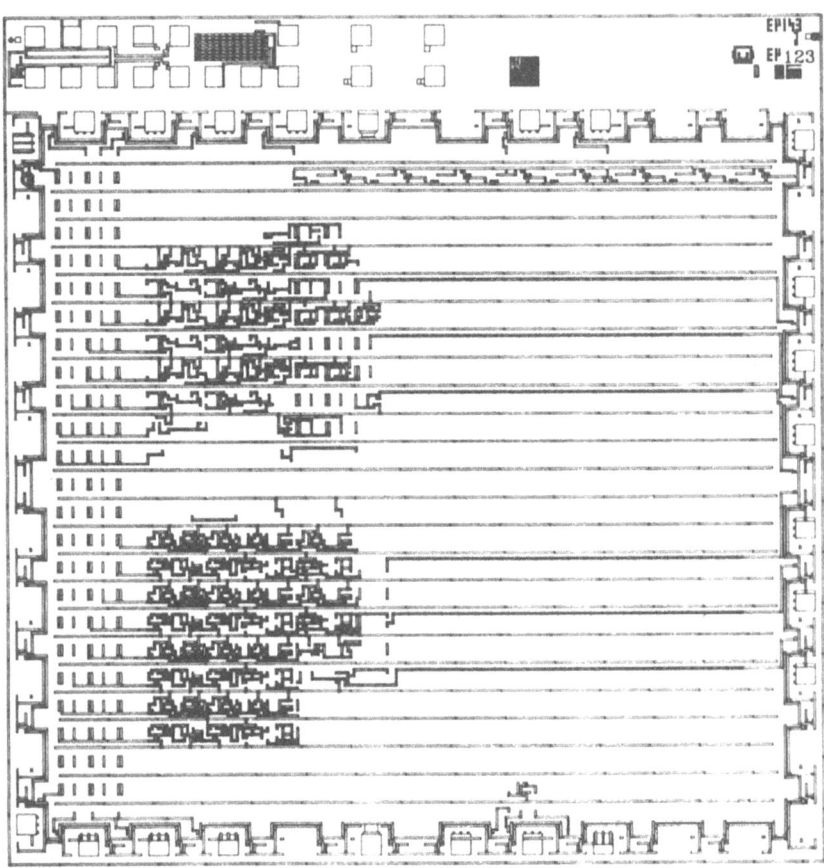

Figure 8.2 The ED500 Gate Array, showing two students' designs

The GATEWAY Gate Array Design Exercise   123

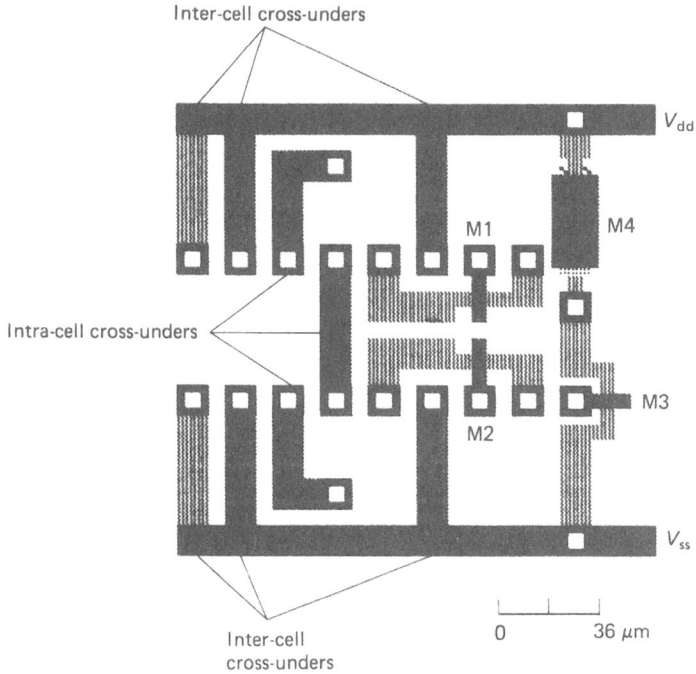

Figure 8.3a The ED500 array cell (layout)

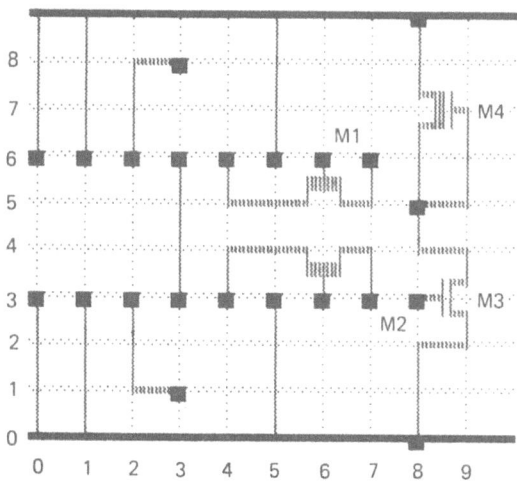

Figure 8.4a The ED500 array cell (schematic)

transistor is *less* than the *drawn length* by $(2 * 1.25)\,\mu m$. The resultant parameters are given in table 8.1a.

The basic logic cell can be configured in many ways to carry out a range of functions.

Table 8.1a ED500 Gate Array Transistor Parameters

Parameter	Transistors M1, M2, M3	Transistor M4
Type	Enhancement	Depletion
Drawn W	$6\,\mu m$	$6\,\mu m$
Drawn L	$6\,\mu m$	$36\,\mu m$
Actual W	$6\,\mu m$	$6\,\mu m$
Actual L	$3.5\,\mu m$	$33.5\,\mu m$
Carrier mobility	$200\,cm^2/V\,s$	$600\,cm^2/V\,s$
W/L (actual)	1.7	0.18
K'	$30\,\mu A/V^2$	$25\,\mu A/V^2$
β	$51.42\,\mu A/V^2$	$4.48\,\mu A/V^2$

8.3.1 The 'Hardwired' Inverter M3/M4

The inverter circuit consists of enhancement transistor M3 and depletion transistor M4. Its input and output points are at grid positions (8,3) and (8,5) respectively.

Table 8.1a shows that β_e/β_d for the inverter is 11.484. This is much greater than the required value of 4 (see equation (2.12)) so an inverter created from one of the enhancement transistors together with M4 would certainly have an adequately low logic '0' output level. Figure 8.5a shows a simplified timing diagram for a logic pulse passing through the inverter structure. The operation of the nMOS inverter has already been dealt with in chapter 2, so only a summary is given here.

(1) Initially the gate of M3 (8,3) is at logic '0', M3 is OFF and the output of the Inverter (8, 5) is therefore at 5V (logic '1').
(2) As the input voltage on the gate of transistor M3 (at grid position 8,3) rises to 5V, M3 turns ON and current flows through M3 and M4. The β_e/β_d value ensures that the output voltage of the inverter (at 8,5) falls to a low enough level to give a logic '0' with an adequate noise margin.
(3) The input to grid position (8,3) returns to logic '0', M3 is turned OFF, and the 'resistive pull-up' effect of M4 returns the output at grid position (8,5) to logic '1'.

Notice that the inverter rise time is illustrated as being longer than the fall time. It will actually be about ten times longer; further details are given in section 8.3.3.

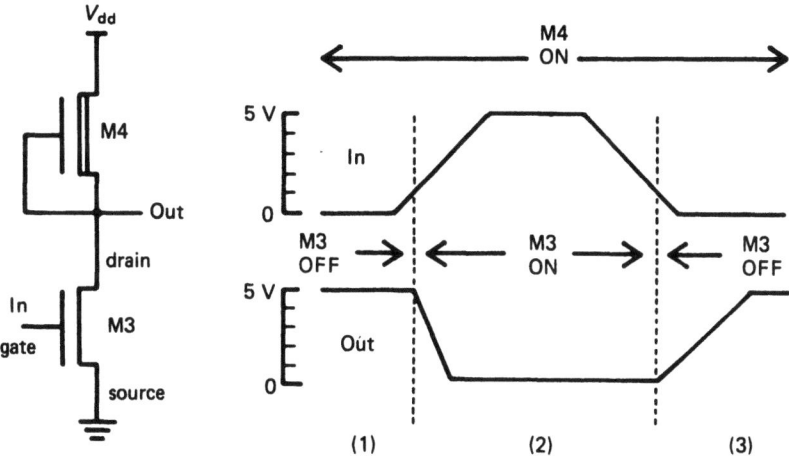

Figure 8.5a Operation of an ED500 inverter

8.3.2 Use of Transistors M1 and M2

Transistors M1 and M2 are 'isolated' enhancement mode devices that can be used along with M3, or as replacements for M3, in logic elements such as NOR or NAND gates.

The gate of transistor M1 is at grid position (6,6) with the source and drain (which are interchangeable) at (4,6) and (7,6). For transistor M2 the gate is at grid position (6,3) with the source and drain (again interchangeable) at (4,3) and (7,3).

To configure the ED500 cell as a logic gate, it will generally be necessary to use the intra-cell 'cross-under' connections present between grid locations (2,3)-(2,1)-(3,1) and (3,3)-(3,6) (note the paraxial definitions here). These 'cross-unders' allow connections to be made from place to place by connecting the signal into and through the underlying silicon structure. This leaves direct tracks free (for example, (0,7)-(9,7) and (0,2)-(9,2) and (0,4)-(9,4)) for metal to be routed straight through the cell.

Figure 8.6a shows the ED500 cell configured as a two-input NOR gate. Here the black areas represent the metal interconnections that form the required configuration. M1 and M3 are connected in parallel such that if the gate of *either* M3 *or* M1 (or both) is at logic '1', there will be a low resistance path from (8,5) to the 0V supply rail, and the output voltage at grid location (8,5) will be LOW. M2 is unused and is left isolated.

For a three-input NOR gate, all three enhancement mode devices are connected in parallel. Here the cross-unders have to be used to make the required connections, especially to V_{dd} (+5 V) or to V_{ss} (0 V). NOR gates with more than three inputs can be constructed, but more pull-down transistors than are

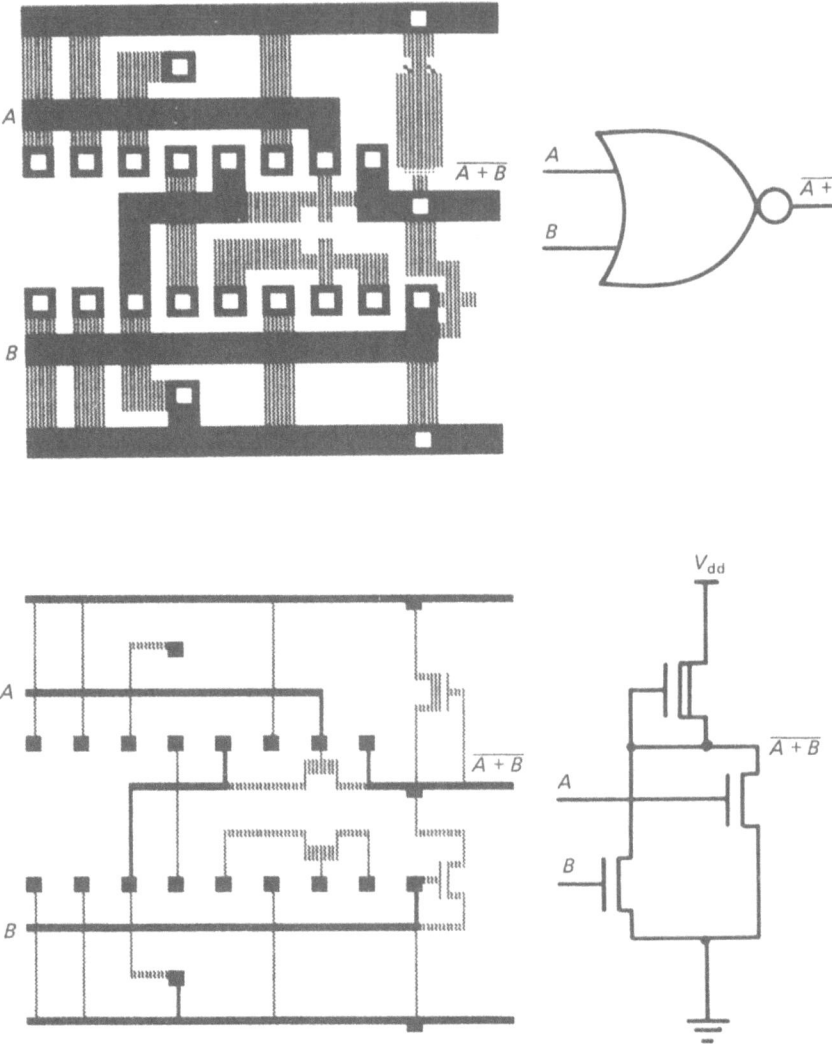

Figure 8.6a   An ED500 NOR gate

available in one cell are required. The extra devices are found by 'stealing' transistors M1 and M2 from another cell.

To realise a two-input NAND gate (figure 8.7a) transistors M1 and M2 are connected in series and transistor M3 is disabled (turned 'OFF') by connecting its gate to V_{ss}. In this circuit configuration, the output will only be 'pulled-down' to logic '0' if M1 AND M2 are both 'ON', yielding the NOT(AND) or NAND function. If a three-input NAND gate is required, one of the M1 or M2 transistors from another cell will have to be 'stolen' and connected in series with M1 and M2 from the 'home' cell. Transistor M3 from the 'home' cell will have to be turned 'OFF' by connecting its gate to V_{ss}. The three enhancement transistors in a three-input NAND gate have a combined 'resistance', when switched 'ON', which is only just low enough to produce a convincing logic '0'. The value of $\beta_{e(combined)}$ divided by β_d is $11.48/3 = 3.83$ which is slightly *less* than the minimum value laid down in section 2.4.1. If we carry out the analysis of this gate we find that the logic '0' level will actually be about $0.11 V_{dd}$, implying that the noise margin has been reduced slightly from the normal value of $0.1 V_{dd}$ to $0.09 V_{dd}$. The three-input NAND gate is only marginally acceptable and should be avoided whenever possible.

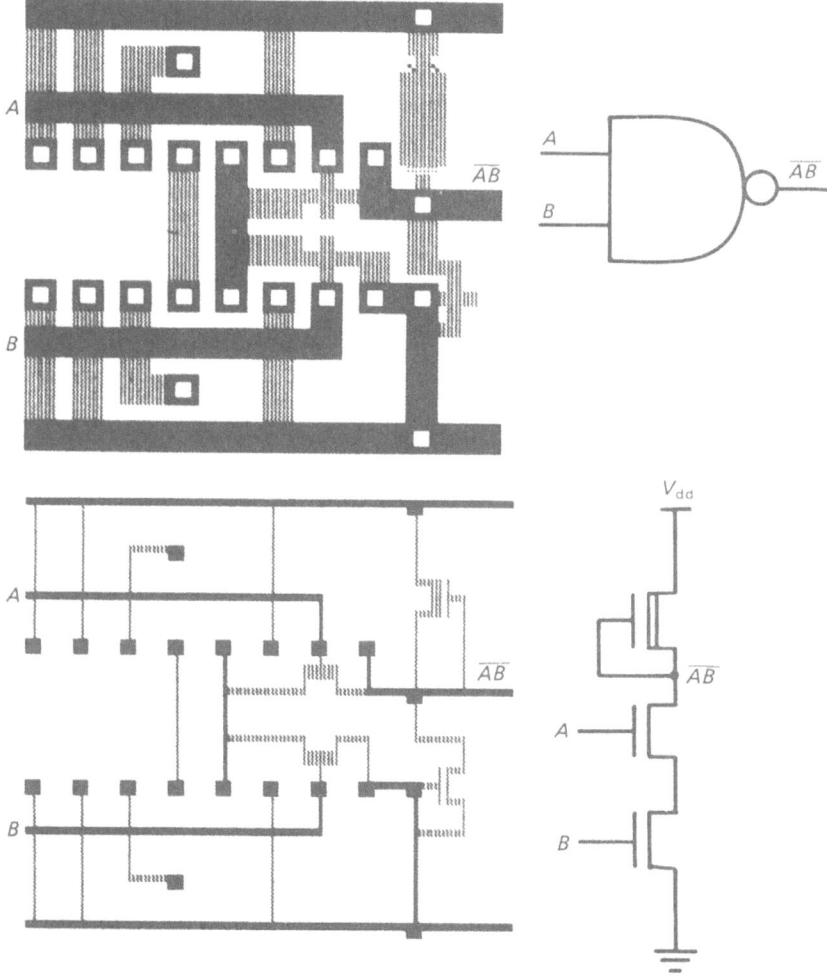

Figure 8.7a An ED500 NAND gate

A four (or more)-input NAND gate would have an even lower noise margin and would not function reliably.

Complex gates can often be realised in a single ED500 cell. The function performed by the configuration shown in figure 8.8a is OUTPUT=$\overline{(AB+C)}$. This example should be examined carefully, to ensure that both the logical function performed and the topology of the interconnections are fully understood.

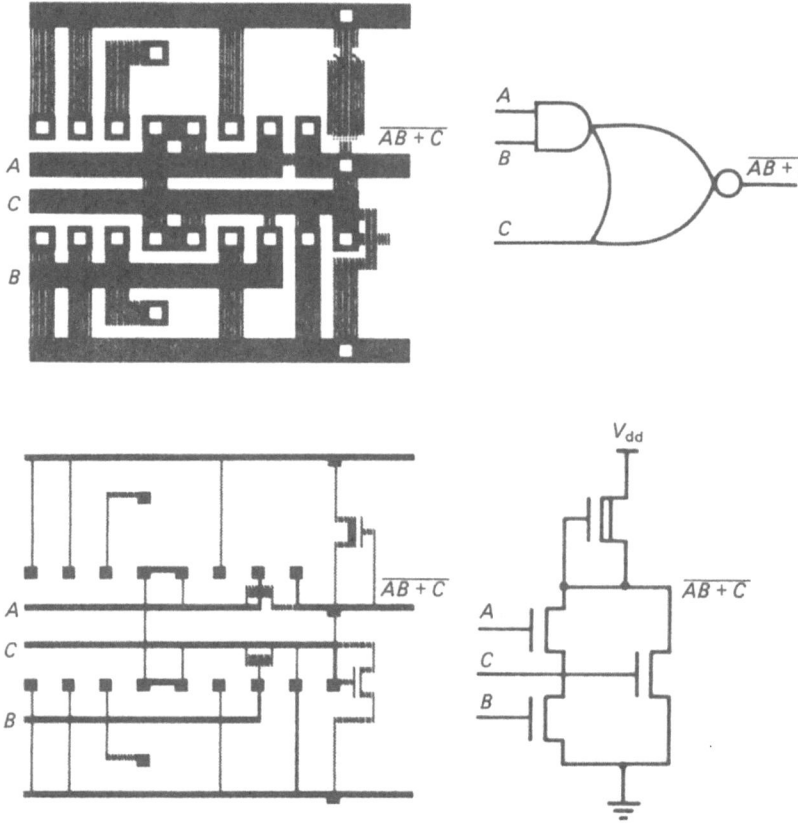

Figure 8.8a An ED500 NOR(AND) gate

Figure 8.9a shows a 4 × 4 array of cells. This diagram can be used, with tracing paper, to perform the GATEWAY exercise without the CAD tools. Although the cells repeat perfectly along the array in the x-direction, they are alternately 'mirror-imaged' between adjacent rows in the y-direction. This minimises the number of V_{dd} and V_{ss} lines that run through the array since adjacent rows 'share' power supply lines. It means, however, that on odd-numbered rows the cells are 'upside down'.

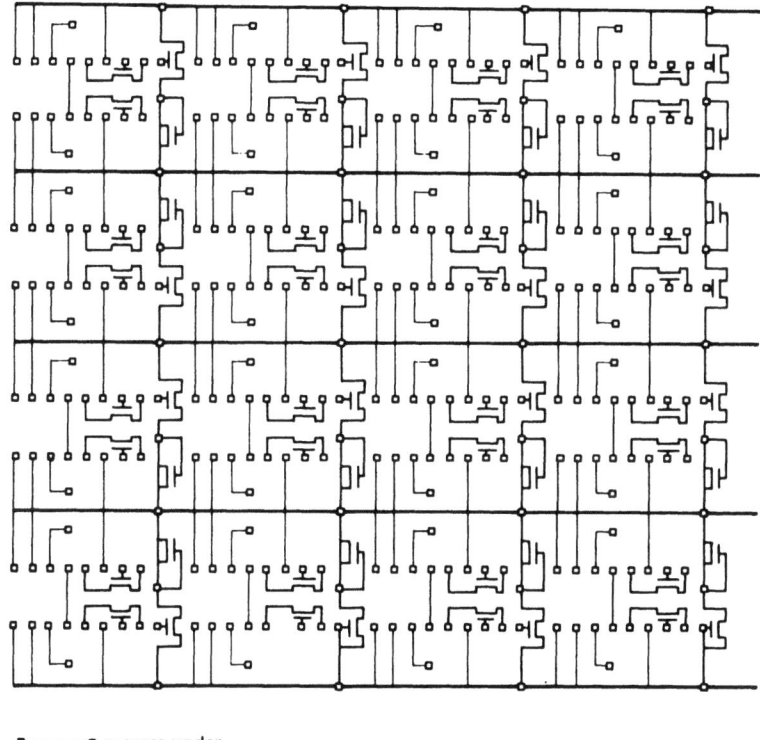

○────○ = cross-under

Figure 8.9a A 4 * 4 cell section of the ED500 gate array

Figure 8.9a also emphasises the usefulness of the cross-unders that occur in each cell at grid locations (0,3), (0,6); (1,3), (1,6); (5,3), (5,6). These permit connections *between* cells in the y-direction *under* the (metal) V_{dd} and V_{ss} lines. The metal power supply lines are pre-defined and cannot be reconfigured in the exercise. The only way to run connections in the y-direction is to exploit these inter-cell cross-unders.

8.3.3 ED500 Gate Array Performance

The overall emphasis of the exercise has to be on exploring the problems of logic design and interconnection, but in chip design in general the ability to achieve a desired performance is of paramount importance. For this reason we must investigate the speed of operation of the logic gates that we can construct using the ED500 design.

The ED500 cell output has a speed characteristic that depends on the capacitive loading of the output node such that the rise time of the node is

$$t_{rise} \approx 500 * C \text{ (ns)} \qquad (8.1a)$$

where C is the total capacitive loading on the output (measured in picoFarads). These figures are extremely conservative (that is, they overestimate the delay times). Use of equation (2.13) gives a value for the rise time of about 320 ns per pico-Farad of loading, so there is a generous margin for error. This helps to ensure a 'safe' design, as is usually good practice. The rise time for all types of gates will be approximately equal (given an equal load capacitance, C) as all the pull-up transistors have equal aspect ratios.

The fall time for an inverter can be shown to be approximately $t_{\text{rise}}/(\beta_e/\beta_d)$ and thus the inverter fall time, t_{fall}, can be taken to be

$$t_{\text{fall}} \approx 50 * C \text{ (ns)} \tag{8.2a}$$

Once again there is a generous margin for error.

The fall time for NOR gates should be approximately equal to that of the inverter, but for a two-input or three-input NAND gate the fall time should be multiplied by 2 or 3 respectively.

Equations (8.1a) and (8.2a) gave rise and fall times in terms of the inverter load capacitance. In order to calculate this capacitance we need to know the capacitances of the metal connections, the cross-under interconnections and transistor gates. These are listed in table 8.2a and a typical nodal capacitance calculation is discussed later in this chapter.

Table 8.2a Capacitances of Gate Array Features

Feature	Capacitance	Unit
Metal	0.006	pF/grid
Gate inputs	0.02	pF(each)
Inter-cell cross-under	0.1	pF(each)
Intra-cell cross-under	0.05	pF(each)

8.4 The ED500C Gate Array

The gate array to be used is a silicon gate (CMOS) design (device number ED500C) containing a total of 552 logic cells (24 × 23). The logical performance of the chip is determined by how these cells are configured and interconnected. The ED500C also contains 40 peripheral cells which can be configured as either input or output bonding pads. Each ED500C chip will support two separate designs with common data inputs, and is fabricated on a process which has a minimum feature size of 3 μm. The single cell of the device, shown in figure 8.3b, contains a total of four MOS transistors. Figure 8.3b also shows some pre-defined interconnections or underpasses within the cell. These interconnections, which may be thought of as 'subways', exist deep within the gate array structure as sections of diffusion or polysilicon and do not impede any metal

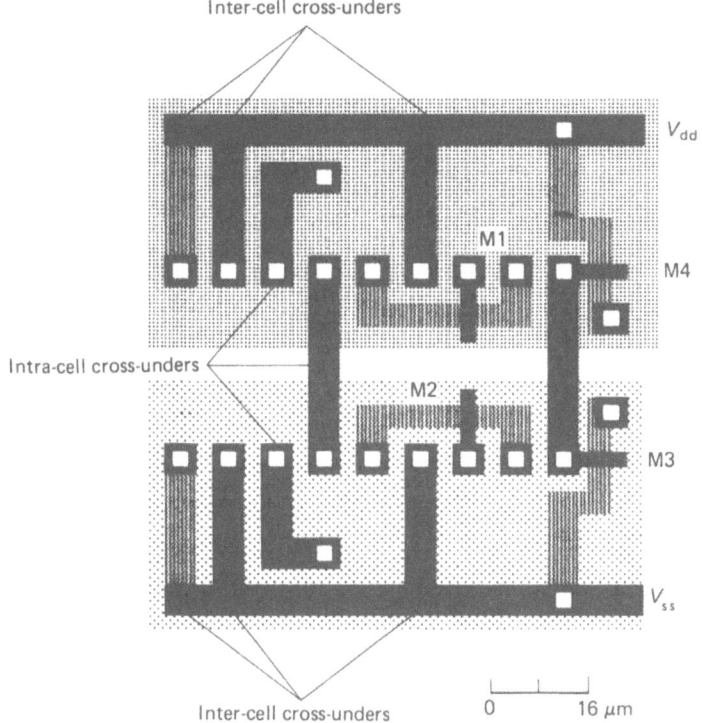

Figure 8.3b The ED500C array cell (layout)

tracks ('highways') which cross over them. The power supply rails (0V and 5V) exist as two horizontal tracks at the top and bottom of the cell.

In figure 8.4b the basic ED500C gate array cell has been redrawn in schematic form with a 9 μm grid to identify the various features to which connections can be made. The grid is labelled with indices '0' through '9' in the x-direction and indices '0' through '9' in the y-direction. This coordinate system will be used throughout the following discussions. Note that all connections must be made by paraxial lines (that is, along the dotted grid lines) and that *no angled lines can be permitted with the ED500C*. This restriction is common in ASIC design, as it simplifies automated layout, and it does not significantly impede layout efficiency. (The one area of ASIC design where non-orthogonal lines are *de rigueur* is in memory design.)

Estimates of likely chip performance must take account of the *lateral underdiffusion*. For the process used this is 0.45 μm so the *actual length* of each transistor is *less* than the *drawn length* by $(2 * 0.45)$ μm. The resultant transistor parameters are given in table 8.1b.

Integrated Circuit Design

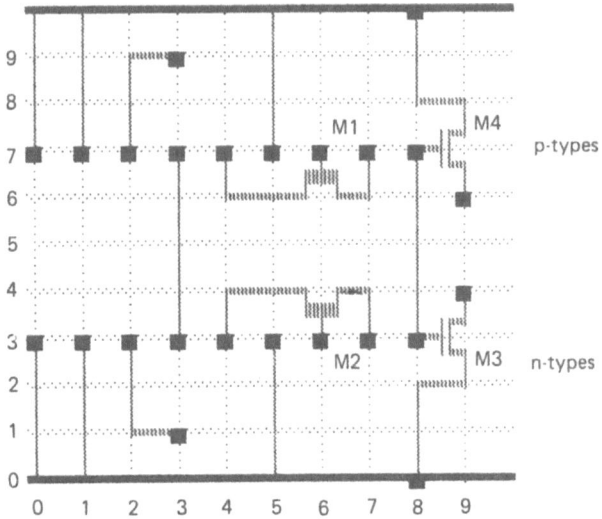

Figure 8.4b The ED500C array cell (schematic)

The basic logic cell can be configured in many ways to carry out a range of functions.

Table 8.1b ED500C Gate Array Transistor Parameters

Parameter	Transistors M1, M3	Transistors M2, M4
Type	n-enhancement	p-enhancement
W	4.5 μm	4.5 μm
Drawn L	3 μm	3 μm
Actual L	2.1 μm	2.1 μm
W/L (actual)	2.1	2.1
K'	40 μA/V^2	15 μA/V^2
β	84 μA/V^2	31.5 μA/V^2

8.4.1 The 'Hardwired' Inverter M3/M4

The inverter circuit consists of n-type enhancement transistor M3 and p-channel enhancement transistor M4. The inverter 'output' is not fully connected to allow the design of logic gates, but can be created by connecting grid points (9,4) and (9,6) together by a strip of metal. Its input and output points are then at grid positions (8,3) or (8,7) and (9,4)–(9,6) respectively.

Table 8.1b shows that while the transistor aspect ratios are equal, the parameters β are not, owing to the disparity between the electron and hole mobilities, and their concomitant process gain factors. Figure 8.5b shows a simplified timing diagram for a logic pulse passing through the inverter structure. The operation of the CMOS inverter has already been dealt with in chapter 2, so only a summary is given here.

(1) Initially the gates of M3 and M4 at (8,3) and (8,7) are at logic '0', M3 is OFF, M4 is ON, and the output of the inverter (9,4-9,6) is therefore at 5V (logic '1').
(2) As the input voltage on the transistor gates rises to 5V, M3 turns ON, M4 turns OFF, and the output voltage of the inverter (at 9,4-9,6) falls to a logic '0'.
(3) The input to grid positions (8,3 and 8,7) returns to logic '0', M3 is turned OFF, M4 is turned ON, and the output at grid positions (9,4-9,6) returns to logic '1'.

Notice that the inverter rise time is illustrated as being longer than the fall time. It will actually be about twice as long; further details are given in section 8.4.3.

8.4.2 Use of Transistors M1 and M2

Transistors M1 and M2 are 'isolated' enhancement mode devices that can be used along with M3 and M4, or as replacements for M3 and M4, in logic elements such as NOR or NAND gates.

The gate of transistor M2 is at grid position (6,3) with the source and drain (which are interchangeable) at (4,3) and (7,3). For transistor M1 the gate is at grid position (6,7) with the source and drain (again interchangeable) at (4,7) and (7,7).

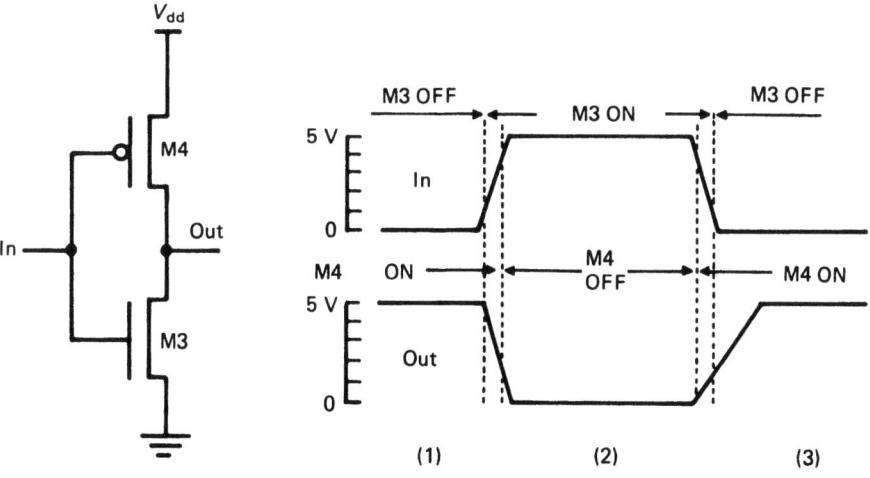

Figure 8.5b Operation of an ED500C inverter

134 Integrated Circuit Design

To configure the ED500C cell as a logic gate, it will generally be necessary to use the intra-cell 'cross-under' connections present between grid locations (2,3)-(2,1)-(3,1), (2,7-2,9-3,9) and (3,3)-(3,7) (note the paraxial definitions here). These 'cross-unders' allow connections to be made from place to place by connecting the signal into and through the underlying silicon structure. This leaves direct tracks free (for example, (0,8)-(9,8) and (0,2)-(9,2) and (0,4)-(9,4)) for metal to be routed straight through the cell.

Figure 8.6b shows the ED500C cell configured as a two-input NOR gate. Here the black areas represent the metal interconnections that form the required

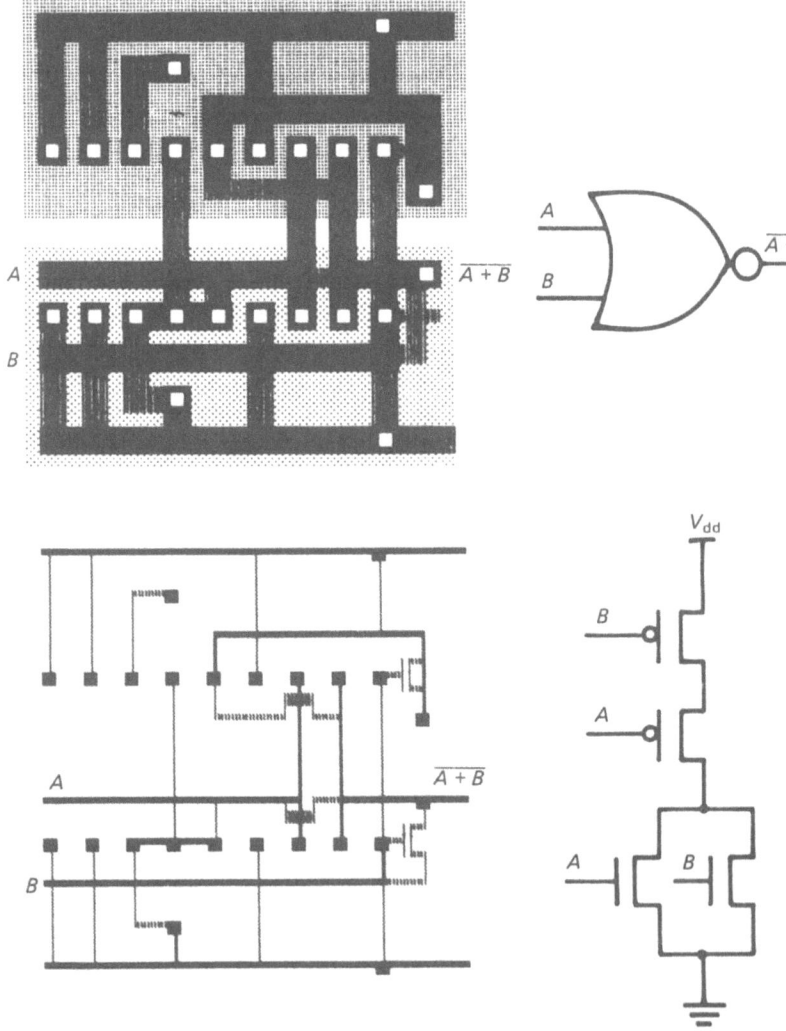

Figure 8.6b An ED500C NOR gate

configuration. M1 and M3 are connected in parallel such that if the gate of *either* M3 *or* M1 (or both) is at logic '1', there will be a low resistance path from (9,4) to the 0 V supply rail, and the output voltage at grid location (9,4) will be LOW. M2 and M4 are connected in series such that *only* if both of their gates are at a logical '0' (that is, *neither A NOR B* are a logical '1') will the output be a logical '1'.

For a three-input NOR gate, three n-channel enhancement mode devices are connected in parallel and three p-channel devices in series. Here the cross-unders have to be used to make the required connections, especially to V_{dd} (+5 V) or to V_{ss} (0 V) and more than one gate array cell will be utilised. The extra devices are found by 'stealing' transistors M1 and M2 from another cell.

To realise a two-input NAND gate (figure 8.7b) transistors M1 and M2 are connected in series and transistors M3 and M4 in parallel. In this circuit configuration, the output will only be 'pulled-down' to logic '0' if M1 and M2 are both 'ON', yielding the NOT(AND) or NAND function. If a three-input NAND gate is required, transistors from another cell will have to be 'stolen' as in the case of the three-input NOR. There is no strong preference in CMOS for NOR logic as there is in nMOS, since NOR logic gives a series 'pull-up tree' of transistors and NAND logic gives a series 'pull-down tree'. The noise margins are the same for both cases, and there is not a significant speed advantage in either form. The NAND gate is easily derived by interchanging the types of the transistors in the NOR gate and interchanging the 0 V and 5 V supply rails. This illustrates the elegant symmetry of CMOS.

Complex gates cannot be realised in a single ED500C cell. The function performed by the configuration shown in figure 8.8b is OUTPUT=$\overline{(AB+C)}$. This example should be examined carefully, to ensure that both the logical function performed and the topology of the interconnections are fully understood.

Figure 8.9b shows a 4 × 4 array of cells. This diagram can be used, with tracing paper, to perform the GATEWAY exercise without the CAD tools. Although the cells repeat perfectly along the array in the *x*-direction, they are alternately 'mirror-imaged' between adjacent rows in the *y*-direction. This minimises the number of V_{dd} and V_{ss} lines that run through the array since adjacent rows 'share' power supply lines. It means, however, that on odd-numbered rows the cells are 'upside down'.

Figure 8.9b also emphasises the usefulness of the cross-unders that occur in each cell at grid locations (0,3), (0,7); (1,3), (1,7); (5,3), (5,7). These permit connections *between* cells in the *y*-direction *under* the (metal) V_{dd} and V_{ss} lines. The metal power supply lines are pre-defined and cannot be reconfigured in the exercise. The only way to run connections in the *y*-direction is to exploit these inter-cell cross-unders.

8.4.3 ED500C Gate Array Performance

The overall emphasis of the exercise has to be on exploring the problems of logic design and interconnection, but in chip design in general the ability to achieve a

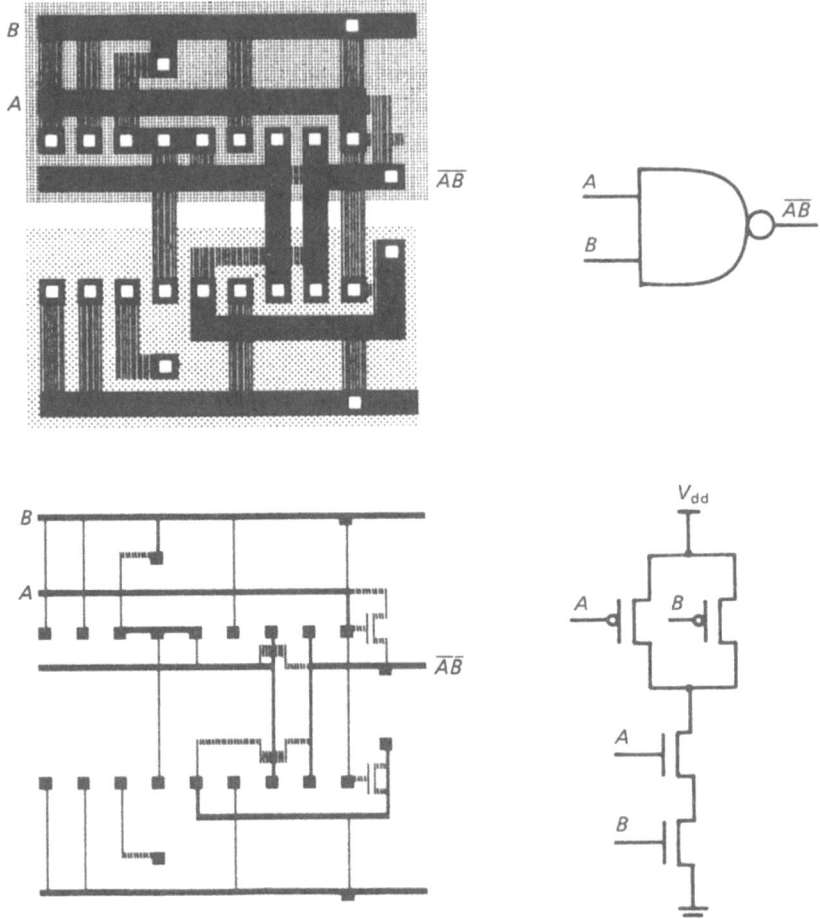

Figure 8.7b An ED500C NAND gate

desired performance is of paramount importance. For this reason we must investigate the speed of operation of the logic gates that we can construct using the ED500C design.

The ED500C cell output has a speed characteristic that depends on the capacitive loading of the output node such that the rise time of the node is

$$t_{\text{rise}} \approx 30 * C \text{ (ns)} \tag{8.1b}$$

where C is the total capacitive loading on the output (measured in picoFarads). These figures are conservative (that is, they overestimate the delay times). Simulation results show that the rise time is about 25 ns per pico-Farad of

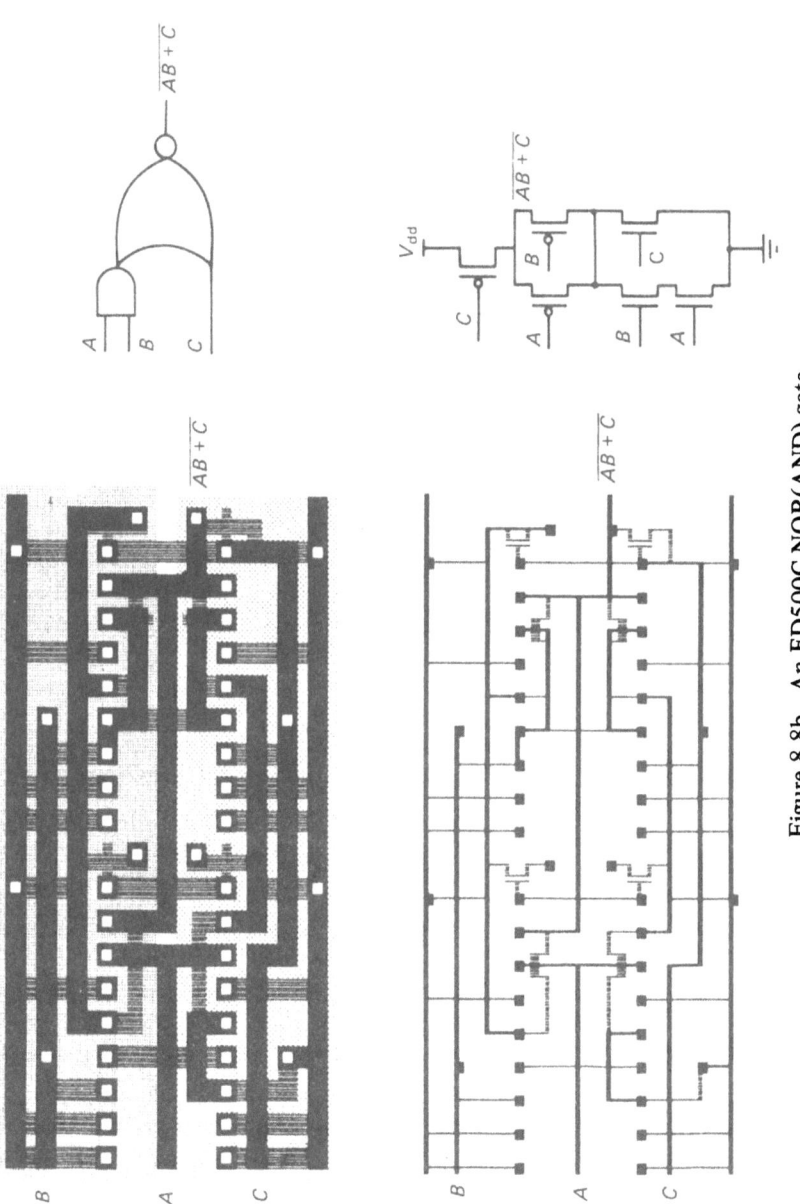

Figure 8.8b An ED500C NOR(AND) gate

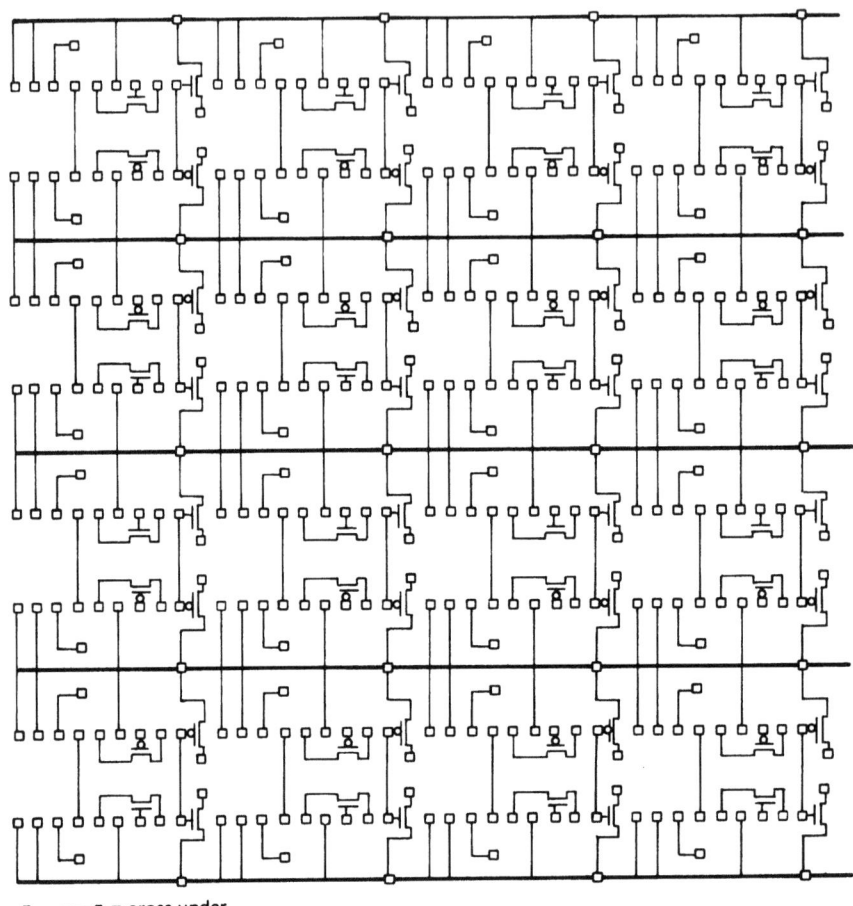

o———o = cross-under

Figure 8.9b A 4 * 4 cell section of the ED500C gate array

loading, so there is a margin for error. This helps to ensure a 'safe' design, as is usually good practice.

The fall time for an inverter might be expected to be approximately $t_{rise}/(\beta_n/\beta_p)$, which would represent a factor of almost 3, as dictated by the different electron and hole mobilities. In reality, effects such as the difference beween n-type and p-type doping levels in the silicon substrate reduce the disparity between the speed of n-type and p-type transistors to a factor of 2. For the ED500C, the inverter fall time, t_{fall}, can therefore be taken to be

$$t_{fall} \approx 15 * C \text{ (ns)} \tag{8.2b}$$

Once again there is a generous margin for error.

The fall time for NOR gates should be approximately equal to that of the inverter, but for a two-input or three-input NAND gate the fall time should be multiplied by 2 or 3 respectively. Similarly, the rise time for an inverter should be multiplied by 2 or 3 to give the rise time for a two-input or three-input NOR gate.

Equations (8.1b) and (8.2b) gave rise and fall times in terms of the inverter load capacitance. In order to calculate this capacitance we need to know the capacitances of the metal connections, the cross-under interconnections and transistor gates. These are listed in table 8.2b and a typical nodal capacitance calculation is discussed in the next section.

Table 8.2b Capacitances of Gate Array Features

Feature	Capacitance	Unit
Metal	0.003	pF/grid
Gate inputs	0.01	pF(each)
Inter-cell cross-under	0.04	pF(each)
Intra-cell cross-under	0.02	pF(each)

8.5 ED500/ED500C Speed Performance Calculations

As the method of calculating the accumulated delay through a series of gates is common to nMOS and CMOS technology, this discussion is relevant to both the ED500 and ED500C arrays. Figure 8.10 shows a typical section of ED500(C) circuitry. Here, the output of the NOR gate passes through several inter-cell and intra-cell underpasses, and through a length of aluminium track. Finally, it drives two input transistor gate capacitances. All these capacitances are in parallel, and are therefore summed to give the total load capacitance. This value is then sub-

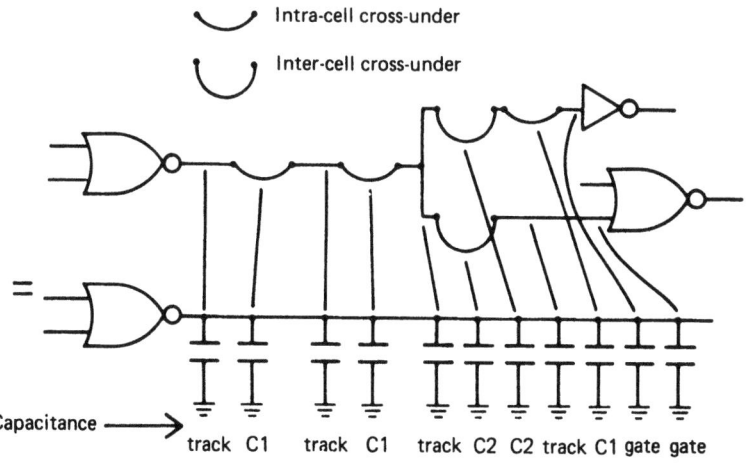

Figure 8.10 A specimen timing calculation (ED500 and ED500C)

stituted into equations (8.1) and (8.2) to give estimates of the rise and fall times of the signal coming from the logic gate *on whose output the load capacitance is present*. This calculation relates only to the output of the gate for which we calculated the load capacitance, and does not tell us anything about the delay through any other gates to whose inputs this output is connected.

Using the capacitances values given in table 8.2a or 8.2b, the total capacitance of a given interconnection (or circuit 'node') can be calculated and used to estimate the circuit delays for critical parts of the design. Referring to figure 8.10, the capacitance for use in rise/fall time calculations would be

$$C(\text{Total}) = 3C1 + 2C2 + 2CG + nCM \tag{8.3}$$

where n depends on how many grid point's worth of track capacitance have been included. To calculate the performance of the entire chip, the slowest portions should be identified and examined, allowing the 'critical' signal paths to be found. Normally, it is not necessary to carry out a delay calculation for all signal paths. The detailed values of $C1$, $C2$, etc. can be calculated from tables 8.2a and 8.2b.

It is not accurate to sum all the capacitances along a signal path regardless of intervening gates/inverters and use this in an overall calculation of circuit speed. Some nodes will be rising and others will be falling. The *delays* through each gate in that signal path should be calculated *individually* and summed, to give the 'worst-case' delay. Clearly, the functionality of the circuit should also be examined (for example, an inverter's input and output cannot be rising simultaneously!). Equally clearly, tracks must not run over unused cell outputs, as this could cause logic errors.

The significant capacitance of transistor gates dictates that these should remain unconnected if unused. This is not an unbreakable rule, and it is sometimes necessary to run a track over an unused feature. The load capacitance on the gate driving this track will be increased and the resulting degradation in its performance should be calculated.

Good use of circuit macros makes timing calculations much less painful, as the same calculation often suffices for several areas of the array.

Delay calculations should *not* be left until the end of a design, as this may result in time being wasted laying out a circuit that cannot meet the timing specification. It is far better to keep an eye on the delays of elements *as they are designed*, minimising unnecessary capacitances wherever possible.

8.6 Design Methods, Discipline and the GATEWAY CAD

If it is desired to implement a complex function which has been specified at a high level, it is not sufficient to be able to design merely NOR and NAND gates. A structured and disciplined design style is of great importance if errors are to be avoided (or at least detected and corrected at an early stage). In addition, an inefficient approach to design will result in wasted time. This is undesirable in a student exercise, and potentially disastrous in an industrial context.

Chapter 7 outlined the principles behind good chip design practice and gave a suggested 'design trajectory'. This will take the designer efficiently from the specification of his chip's required function to gate array interconnect patterns in such a way that he will have a high degree of confidence that his design will function as intended.

The GATEWAY CAD has been developed with this design practice firmly in mind, and will assist designers in adhering to it. Discussion of the GATEWAY CAD is omitted however, as it is intended that the exercise be of value without the associated software.

8.7 GATEWAY Project Assignment

The following assignment is designed to be given to students as the GATEWAY exercise at the beginning of the academic year. This work can run concurrently with other coursework — that is, over the entire (active) academic year.

You are working for a large, international company whose UK installation is concerned with the design and manufacture of computer peripherals (terminals, plotters, disc drives, etc.). A new series of graphics terminals is under development, using a combination of standard parts (microprocessors, memory, etc.) where appropriate, and Application Specific Integrated Circuits where the gain to be made by using customised chips is substantial. At the end of the project, fifty prototype terminals will be assembled, so all parts have to be available and tested by that time.

8.7.1 Circuit Description and Details

You, as a junior engineer, with aspirations to progress further, have been given responsibility for committing part of the terminal series' logic circuitry to gate array. Figure 8.11 shows the function you are expected to realise.

Two of your gate arrays will be used to monitor the screen databus (one for each of the x and y coordinates) to detect whether or not a data value on the bus lies within the limits of the screen. Each gate array must therefore produce three outputs showing whether its coordinate datum is within the screen ('PLOT'), greater than the upper limit ('GT') or below the lower limit ('LT'). These signals will be used to avoid time being wasted by an attempt to plot points which are not in the display area.

The gate array will be permanently configured as *either* an 'x' or a 'y' model. The data on the d0-7 lines represents *either* the x or y coordinate of the pixel to be illuminated, while the o0-7 data are corresponding offsets. Thus an x-gate array and a y-gate array will take a total of four inputs: X, X_{offset}, Y and Y_{offset}, and the actual pixel illuminated will have coordinates $(X + X_{\text{offset}}, Y + Y_{\text{offset}})$.

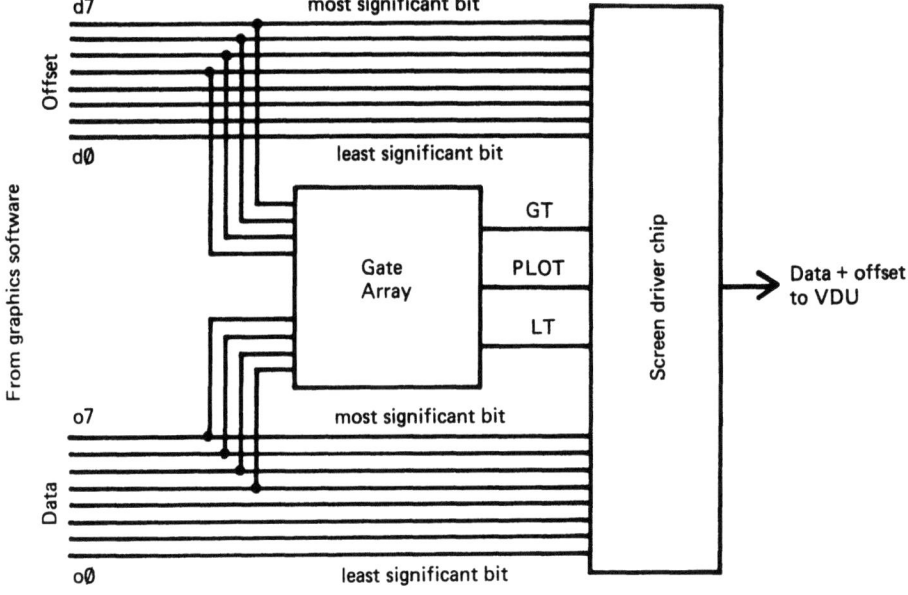

GT   = (Data + offset) > Upper limit
LT   = (Data + offset) < Lower limit
PLOT = (Lower limit) ≤ (Data + offset) ≤ (Upper limit)

Figure 8.11 The GATEWAY project assignment (schematic)

It has been decided that it is sufficient to examine only the four most significant bits of the data and offset signals, and to tolerate the slight loss of precision that this entails.

The gate array will therefore be expected to ADD the upper four bits of the 'data' and 'offset' inputs, and subsequently to compare the upper four bits of the resultant sum with the upper and lower limits corresponding to the screen dimensions. These limits will be coded into the chip by the designer (you) and should NOT be regarded as inputs from the outside. The only uncertainty in the design specification is in these limits, whose final values in the prototype terminal cannot be known until the screen dimensions are decided. This decision will be made during February. Furthermore, the same basic gate array design is to be usable as both x and y data monitors, so the limits must be easily changeable, as the screen is not square. You will therefore have to make the limits 'mask programmable' such that only a few well-documented changes need be made to the design to change the limits. Careful use of macros can aid this part of the exercise. For the prototype terminals, two gate arrays will be used, so a total of 100 of your devices will be required. You will only be required to produce one mask, with a detailed and concise set of instructions to allow its limits to be changed.

The GATEWAY Gate Array Design Exercise

Clearly, a well-structured design is important here, but compactness of design is also important. While you should not 'squash' your logic into the barest minimum of cells, at the expense of design structure, you should attempt to minimise the number of cells used. As a guide, 30–60 cells is a typical GATEWAY project.

The system designer has assumed a data rate of 100 KHz (400 KHz for the CMOS exercise), and that your gate array will have a maximum input/output delay of less than 5 μs (1.25 μs for the CMOS exercise). You should be able to provide calculated evidence that your design will meet this performance specification. You should, as stated earlier, ensure that this specification is met *AS YOU ARE DOING THE DESIGN*, rather than at the end of the exercise.

8.7.2 Project Management

A bar-graph, showing the timescales for various phases of the graphic terminal prototyping is shown in figure 8.12. You will be expected to follow this schedule. There are three major phases in the project, each of which is associated with a check point on completion. This implies a set of requirements on the status of the work, and also some documentation. The phases are as follows.

Investigation phase

The investigation phase of a new product's lifecycle is the initial study of a proposed development, intended to determine its feasibility, marketability and profitability before entering the formal and more costly full design and production phases. During this phase, the detailed definition of the product will be developed to the stage of projected detailed specification. It is an important

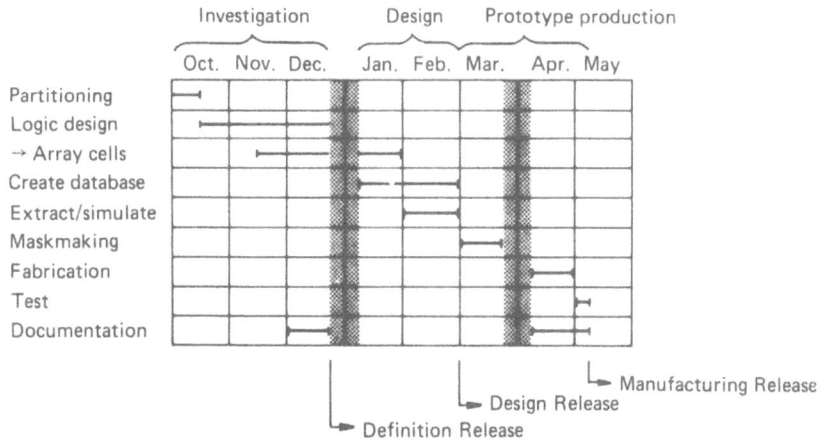

Figure 8.12 Bar-graph, showing timescales for phases of graphic terminal

phase, as it is the point at which silly ideas should be discarded, and good ones given a high priority for further work.

The product with which you are involved has been proposed already, and you are required to carry out the investigation and definition of the gate array section of the VDU control circuitry. By the first check-point this must result in a more-or-less complete gate level description of the function shown in figure 8.11, as only by going this far down the design path can you determine its feasibility as a gate array. In addition, you should begin to think about the implementation of the gates you are using as ED500(C) cells. You should therefore aim, by the end of the investigation phase, to have partitioned your top-level description (figure 8.11) into basic blocks, and to have described the low-level gates as transistors (on paper). You should document this stage in your design in such a way that, firstly, the requirements of the Definition Release are satisfied and, secondly, the work done at this stage is usable directly in the final documentation. The first check-point is marked by the writing of the definition release, which should contain the following information:

(1) A detailed specification of projected gate array, including an estimate of the number of gate array cells, and of the input/output delay. A discussion of the feasibility or otherwise of full gate array development. All the logic gates to be used should be defined, and their probable use of array cells discussed.
(2) A project planning sheet. This should consist of a copy of the bar-chart of figure 8.12, with your work so far, and over the next phase, entered in coloured ink. Remember that the second check-point is another immovable date, and if you let your timescale slip you will either miss it or be working excessively hard during February!

Design phase

During this (cost and labour intensive) phase, the product will be completely designed to the specification determined earlier. The gate array may be designed in terms of gate array cells and then may be entered into a database (using the gate array graphics editor MIDGET). Alternatively, figures 8.9a and 8.9b may be used as the basis of a paper exercise. The design must then be circuit-extracted and logically simulated to prove its correctness. The input/output delay must then be recalculated to take account of the detailed interconnect pattern.

During this time the project will have full status, and will be on its way to the marketplace. At the end of this phase, which is marked by the second check-point and the release, a complete set of design data should be available in a form suitable for passing to the fabricator. The design release is primarily a final product specification.

Prototype production

This phase, the final one in the project, would not, in reality, involve you. It involves the production of a final report which is effectively a manufacturing release, and should include much of the preceding documentation.

8.7.3 Costing of Project

You will be managing your own time as a designer on this project, and must aim to keep within the project budget allocated to you. The overall budget for the gate array development is £6000, to include all design, mask-making, fabrication, bonding and testing costs. Computing costs are absorbed into the charge-rate for design time as an overhead. A local company will make the masks for £2800, and they will be sent to the United States for fabrication by your own company's gate array facility. This will be charged to the project at a rate of £25 per device for $<$ 1000 devices, reducing to £15 per device for larger quantities.

Your own time as a junior designer will be charged to the project at £125/(7-hour)man-day, including overheads (rent of premises, directors' and support staff's salaries, computing time, etc.). Overheads typically amount to 60 per cent of the charge-out rate, so your salary is not the £30 000 that you hoped it was!

Appendix: Inverter Output Rise Time

The inverter output logic '0' level is taken to be $0.1V_{dd}$. When the inverter output, V_{out}, is greater than this value and is rising but is still less than $0.2V_{dd}$, transistor d1 will be *saturated*. Thus for $(0 < t < t_1)$ we should use the current equation (2.6) which gives

$$I_{ds} = \left[\frac{\beta_d}{2}\right] [(0 - (-0.8V_{dd}))^2] \tag{A.1}$$

where

$$\beta_d = \left[\frac{W_d}{L_d}\right] K_d \tag{A.2}$$

Now this current must be flowing into capacitor C, the current equation for which is

$$I = C * \frac{dV_{out}}{dt} \tag{A.3}$$

These two equations lead to

$$C \frac{dV_{out}}{dt} = \frac{\beta_d}{2} * 0.64 V_{dd}^2 \tag{A.4}$$

which implies

$$\frac{1}{C} dt = (\beta_d * 0.32 V_{dd}^2)^{-1} dV_{out} \tag{A.5}$$

If this is integrated over the limits $[0$ to $t_1]$ on the left-hand side (LHS) and $[0.1V_{dd}$ to $0.2V_{dd}]$ on the right-hand side (RHS) this gives

$$t_1 = \frac{C}{3.2 * \beta_d * V_{dd}} \tag{A.6}$$

Appendix: Inverter Output Rise Time

Once the inverter output has risen to $0.2V_{dd}$ the pull-up transistor moves into the *unsaturated* region. We therefore use the current equation (2.3) to obtain

$$I_{ds} = \beta_d \left[(0 - (-0.8V_{dd}))(V_{dd} - V_{out}) - \frac{(V_{dd} - V_{out})^2}{2} \right] \quad (A.7)$$

This, together with the capacitor current equation (A.3) gives

$$\frac{1}{C} dt = \left[\frac{1}{\beta_d} \right] \left[0.8V_{dd}(V_{dd} - V_{out}) - \frac{(V_{dd} - V_{out})^2}{2} \right]^{-1} dV_{out} \quad (A.8)$$

The limits of integration are $[t_1$ to $t_2]$ on the LHS and $[0.2V_{dd}$ to $0.9V_{dd}]$ on the RHS. To integrate the RHS the substitution $[x = (V_{dd} - V_{out})]$ should be made, which changes the limits of integration to $[0.8V_{dd}$ to $0.1V_{dd}]$. From this we obtain

$$\int_{t=t_1}^{t=t_2} \frac{1}{C} dt = \frac{1}{\beta_d} \int_{x=0.8V_{dd}}^{x=0.1V_{dd}} \left[\frac{2}{x(x - 1.6V_{dd})} \right] dx \quad (A.9)$$

This gives us

$$t_2 - t_1 = \left[\frac{C}{(0.8 * \beta_d * V_{dd})} \right] \ln \left[\frac{x}{1.6V_{dd} - x} \right]_{x=0.1V_{dd}}^{x=0.8V_{dd}} \quad (A.10)$$

When we substitute the limits for x we get

$$t_2 - t_1 = \left[\frac{C}{(0.8 * \beta_d * V_{dd})} \right] \ln(15) \quad (A.11)$$

We can now find the total time for the inverter output to rise to $0.9V_{dd}$ by summing equations (A.6) and (A.11). The result is

$$t_{rise} = \frac{4 * C}{(\beta_d * V_{dd})} \quad (A.12)$$

While this analysis may seem complicated, it actually rather oversimplifies the problem. Only first order equations are used and a number of important second order effects (such as body effect) are not included. Practical experience suggests that equation (A.12) underestimates the delay by almost a factor of 2, and a more useful estimate of the inverter rise time is given by

$$t_{rise} \approx \frac{7 * C}{(\beta_d * V_{dd})} \quad (A.13)$$

Bibliography

This list indicates where more detailed discussion can be found on selected topics covered by this book.

Fundamentals of Solid-State Physics
Kittel, C. (1986). *Introduction to Solid-State Physics*, 6th edn, Wiley, New York.

Device Physics
Sze, S. M. (1981). *Physics of Semiconductor Devices*, 2nd edn, Wiley, New York.

Fabrication and Integrated Circuit Technology
Glaser, A. B. and Subak-Sharpe, G. E. (1979). *Integrated Circuit Engineering*, Addison-Wesley, New York.

LSI/VLSI Design
Mavor, J., Jack, M. A. and Denyer, P. B. (1983). *Introduction to MOS LSI Design*, Addison-Wesley, Wokingham, Berkshire.
Mead, C. and Conway, L. (1980). *Introduction to VLSI Systems*, Addison-Wesley, Reading, Massachusetts.
Weste, N. and Eshraghian, K. (1985). *Principles of CMOS VLSI Design*, Addison-Wesley, Reading, Massachusetts.

Integrated Circuit Testing and Reliability
Bennets, R. G. (1984). *Design of Testable Logic Circuits*, Addison-Wesley, Wokingham, Berkshire and Reading, Massachusetts.
Cluley, J. C. (1981). *Electronic Equipment Reliability*, 2nd edn, Macmillan, Basingstoke.

Computer Aided Design
There is no real 'bible'. There are useful and interesting articles in *VLSI Design* magazine (originally *Lambda*), published by CMP Publications Ltd (New York), and *Electronic Design Automation* (originally *Silicon Design*), published by EDA Ltd (London).

Index

Acceptor 4
Alignment marks 66
Application-Specific Integrated Circuit 11
Artificial Intelligence 9
ASIC 11
ATPG 114

Bipolar technologies 75
Bipolar transistor operating modes 76
Body effect 20
Boule 44
Buried contacts 69
   defining gate length 72
Butting contacts 69

CAD 100
Calculators 8
Capacitance
   of diffusion 27
   of metal interconnect 27
   of transistor gate 28
Cells 102
Charge carrier 4
Chip layout 110
Circuit extraction 113
Combinational logic 41
Comparison between logic families 84
Complementary error function diffusion 50
Computer Aided Design 100
Computers in vehicles 8
Constant source diffusion 50
Contact printer 47
Control systems 9
Custom design 93

Dark field mask 44
DeMorgan's rule 81
Depletion of semiconductor 15
Depletion transistor 23
Design discipline 100
Design rules 63
Design verification 104
Diffusion 48
Diffusion capacitance 27
Diffusion coefficient 49
Diode Transistor Logic 77
Direct Coupled Transistor Logic 76
Direct-step-on-wafer 48
Discrete component system 10
Donor 4
Dopants 4
Drain 6
Drive-in diffusion 52
Dry oxidation 48
DTL 77

E-beam mask-making 45
ECL 79
ED500 gate array 120
ED500C gate array 130
Electrical design rules 66
Electron beam mask-making 45
Electronics in communications 8
Electrons 4
Emitter Coupled Logic 79
Enhancement nMOS transistor 22
Enhancement pMOS transistor 22
Etch control 65

Fabrication data, generation of 113
Fanout 31

149

Index

Finite state machine 96
Flash 45
Full custom design 93
Future state of microelectronics 9

GaAs 83
Gallium arsenide 83
Gallium phosphide 83
GaP 83
Gate 6
Gate array design 85
Gate level simulation 106
Gates
   CMOS 32
   nMOS 40
GATEWAY
   CMOS gate array performance 135
   CMOS inverter 132
   CMOS NAND gate 135
   CMOS NOR gate 134
   nMOS gate array performance 129
   nMOS inverter 124
   nMOS NAND gate 126
   nMOS NOR gate 125
GATEWAY CAD 140
GATEWAY design discipline 140
GATEWAY design exercise 119
GATEWAY design methods 140
GATEWAY project assignment 141
GATEWAY project costing 145
GATEWAY project management 143
GATEWAY speed performance
   calculations 139
Gaussian diffusion 51
Geometric design rules 63
Graphic editors 111
Groups 102

Handcrafted design 93
Hardware Description Language 106
HDL 106
HEMT 84
Hierarchical design 102
High Electron Mobility Transistor 84
Holes 4

III–V semiconductors 83
I^2L 82
Indium phosphide 83
Information Technology 9
InP 83

Insulators on integrated circuits 48
Integrated Injection Logic 82
Inversion of semiconductor 15
Inverter, CMOS 24
   logic switching 28
   logical performance 24
   performance (summary) 31
   power consumption 25
   speed performance 26
Inverter, nMOS 35
   rise time 38
   rise time calculation 146
   simple analysis 36
Ion implantation 54
   range 56
   straggle 56

Lambda rules 67
Large Scale Integration 6
Lateral under-diffusion 39, 53
Layout of book 13
Light field mask 44
Limited source diffusion 50
Linear region 17
Logic (nMOS) 35
Logic families, comparison 84
Logic gates, speed performance
   (CMOS) 35
Logical performance (nMOS) 39
Low Power TTL 79
Low Pressure Chemical Vapour
   Deposition 57
LPCVD 57
   of polysilicon 58
   of silicon dioxide 58
   of silicon nitride 58
LSI 6

Macros 102
Mandatory features 66
Mask production 44
Mask registration 47, 67
Masterslice design 85
MESFET 84
Metal interconnect capacitance 27
Metallisation 59
Microprocessor-based system 10
Microprocessors 10
Milestones in IC development 6
Moore's law 7
MOS device, full equations 19

Index

MOS operating regions 16–18
MOS switch—descriptive 16
MOS transistor, physical behaviour 14
MOS transistor equations
   linear region 20
   saturation region 20
MOS transistors as resistive switches 31
MOSFET 6
Multi-project wafers 46

NAND gate
   CMOS 32
   nMOS 40
Negative photoresist 46
Noise margin 37
NOR gate
   CMOS 32
   nMOS 40

Optical mask-making 45
Optical projection printer 47
Other IC technologies 75
Oxidation of silicon 48

Partitioning 103
Photolithography 46
Photoresist 45
Photoresist exposure, control of 65
PLA 95
Placement, automatic 111
Positive photoresist 46
Practical diffusion techniques 52
Pre-deposition stage of diffusion process 52
Process example (nMOS) 60
Process-independent design rules 66
   advantages 73
Production of electronic system 10
Programmable Logic Array 95

Radar 4
Radiation hardness 83
Radio 3
Ratioed logic design 38
Registration, of masks 47
Resistor Transistor Logic 76
Resolution on masks 45
Reticle 45
Routing, automatic 111
RTL 76
Run-out 65

Sapphire, silicon on 82
Saturation current, variations in 22
Saturation region 18
Schottky TTL 79
Scribe channel 66
Self-aligned gate process 54, 62
Semiconductor age, the coming of 4
Sideways diffusion 64
Silicon compiler 112
Silicon dioxide as a mask 49
Silicon wafer production 44
Silicon-on-Insulator 75
Silicon-on-Sapphire 82
Simulation 104
   behavioural (high) level 105
   device level 110
   gate level 106
   switch level 108
Small Scale Integration 6
SOI 75
SOS 82
Source 6
SSI 6
SSI systems 10
Standard cell design 90
Step coverage 59
Step-and-repeat 46
'Stuck-at' faults 114
Substrate back-bias 28
Symbolic editors 111
Synchronous logic 41
System-on-a-chip, advantages 12

Technology-Independent Design Rules 67
Test pattern generation 113
   automatic 114
Test pattern validation 115
Testability, design for 116
Testing 113
   exhaustive 114
   functional 115
   structural 115
Three-five (III–V) semiconductors 83
Threshold voltage 16
Top-down design 103
Transistor gate capacitance 28
Transistor Transistor Logic 78
Transistorised computers 7
Triode 3
TTL 78

Valves (tubes) 3
Very Large Scale Integration 7
VLSI 7

Wafer production 44
Wafer Scale Integration 7

Wet oxidation 48
WSI 7

XOR gate
  CMOS 34
  nMOS 40

MIX
Papier aus verantwortungsvollen Quellen
Paper from responsible sources
FSC® C105338

If you have any concerns about our products,
you can contact us on
ProductSafety@springernature.com

In case Publisher is established outside the EU,
the EU authorized representative is:
**Springer Nature Customer Service Center GmbH
Europaplatz 3, 69115 Heidelberg, Germany**

Printed by Libri Plureos GmbH
in Hamburg, Germany